The Cats of Lamu

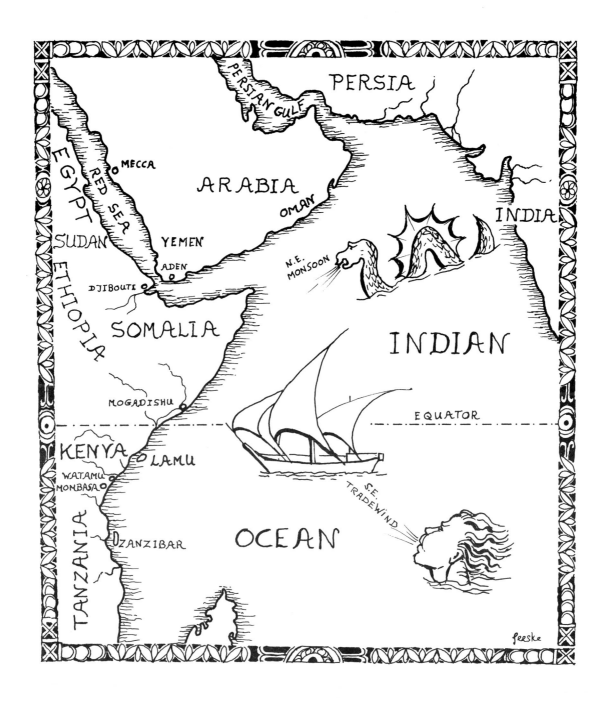

The Cats of Lamu

Text by
JACK COUFFER

Photographs by
Jack and Mike Couffer

THE LYONS PRESS

To the people of Shela,
who opened their doors and their hearts.

The Lyons Press, 31 West 21 Street, New York, NY 10010.
Printed in Spain
D.L. TO : 89 - 1998
Designed by Joel Friedlander, Marin Bookworks
Photographs by Jack and Mike Couffer
10 9 8 7 6 5 4 3 2 1

Library of Congress Cataloging-in-Publication Data
Couffer, Jack.
The cats of Lamu / text by Jack Couffer; photographs by Jack and Mike Couffer.
p.   cm.
Includes bibliographical references.
ISBN 1-55821-675-8
1. Cats—Behavior—Kenya—Lamu Island.   2. Feral cats—Behavior—Kenya—Lamu Island.
3. Cats—Kenya—Lamu Island—Pictorial works.   4. Feral cats—Kenya—Lamu Island—Pictorial works.
5. Lamu Island (Kenya)   6. Couffer, Jack.   I. Title.
SF446.5.C685   1998
599.7'52—dc21   97-43247
CIP

Contents

This book is about some cats I know, but their story can be written only if I also tell about some people—and a place. The place is far away, yet traits of these distinctive cats can be seen in domestic felines anywhere. Many of the characters, both cat and human, could live on anyone's block.

Four or five thousand years ago, African wild cats were domesticated by the Egyptians. Several hundred years and many cat generations later, the wild animal gradually evolved into the domestic cat as we now know it. Today's many varieties of house cats differ noticeably from their Egyptian ancestors, and the cats of present-day Cairo look the same as cats from everywhere else. But not far from Egypt is an island where a unique group of cats resemble those of the ancients. They are remarkable in their social organization and in their place within a human culture. On the Lamu Archipelago, off the coast of Kenya, the cats of the pharaohs may still survive.

One thing is certain: Wherever they originated, Lamu cats have been here for a long time. The towns and ruins of Lamu were thriving trading centers more than a thousand years ago, and archeological digs show that these people kept cats and traded with ports in the Red Sea from the very beginning. Here in Lamu these cats have been genetically isolated by their island environment.

From all that can be determined from the depictions of cats in the tomb paintings and frescoes of ancient Egypt, there appears to have been little variation in color. The cats of the pharaohs were yellow-brown with darker spots or tabby stripes. The colors of Lamu cats, on the other hand, run nearly the entire cat-color spectrum. It is the unique shape of Lamu cats—long legs, slim body, whip tail, short hair, long neck, and small head—that presents a conformation identical to the sacred cats of Egypt.

Painted and sculpted in frescoes and stone, the sacred cats of ancient Egypt—ancestors of all today's house cats—are extinct in their homeland.

Isolated on Lamu Island off the Kenya coast, a group of cats endure that evoke a haunting likeness to those long-gone felines of ancient Egypt. Their long legs, small head, whip tail, short fur, and thin body provide a classic model for an Egyptian sculpture.

Broken bits of pottery that wash in with the tide tell us
that ancient Lamu traded down the Red Sea from Egypt
and across the Indian Ocean from Arabia, the Persian
Gulf, India, and China.

The Lamu cats have fascinated me since my first sight of one more than twenty years ago. I feel today, as I did then, that looking at one is seeing a living incarnation of Egyptian sculpture.

I've observed and cared for Lamu cats since first laying eyes on them and have spent many hours on their home turf watching and photographing them.

I didn't use the modern tools of animal behaviorists for my studies. No radio collars girdled my cats to help me follow their wanderings, and no red or yellow ear tags flagged my subjects. The latter wasn't necessary, as each cat was naturally distinctive in its physical marks. As for pinpointing their wanderings by the radio technology used by field biologists, I did occasionally wish for such sophisticated methods, but it seemed inappropriate for the atmosphere of this place. Anyway, I found out all I needed to know simply by following and watching.

At times I yearned to be able to travel over the walls and onto the roofs with my cats to keep up with their every movement. But that was impossible. Nighttime viewing was also troublesome. As there is no municipal lighting in Lamu town, and only a few globes shine out from windows or vestibules to dimly illuminate the narrow lanes, the streets are dark with pools of shadow. Still, on moonlit nights one can see surprisingly well. The house and garden walls are highly reflective, and shapes or shadows move against them with ample clarity, so the problem of observation is less one of vision than it is of terrain. When a cat disappears over a wall, as inevitably it must, there is usually no way to follow it. Indeed, it would be dangerous to try because Lamuites, like people anywhere, do not look kindly on strangers climbing over their walls.

I attracted notice among the locals for my lurking behavior, and I often felt the cold stab of leery eyes watching me sideways. In the early days of my study, children sometimes followed me around, disrupting my attempt to become a low-profile part of the landscape. They asked me questions—simple ones such as: "What are you doing?" and I found it hard to explain. What could I say? "Watching the cats . . . ?" It made sense to me, but not to them, and only led to more questions. But at last I became known merely as another eccentric, no more odd than many of themselves. Townspeople soon learned what I was doing and called me *Bwana Paka* (Mister Cat)—and accepted me into their fold.

A Walk on the Wild Side

What brought us to Lamu began with a holiday and turned into a lifestyle. We were flying from Nairobi in a chartered plane to a remote hideaway we'd heard about from friends. It was like a dream. The surf below the little Cessna broke in long white lines against mile after mile of deserted beaches. We saw no footprints, no human figures on the sand. Beyond the sea, barrier dunes merged with the gray-green forest that disappeared in haze and the blue smoke of faraway bushfires.

We flew low, only a couple of hundred feet over the

Shela waterfront at high tide.

Shela Beach.

The houses in Lamu are built upon the ruins of earlier civilizations.

shore. Ahead we saw broken-down walls standing on a headland, a pillar of masonry that might have been a tomb, ruins of an ancient town, isolated, lost. No human trail or motor track approached it through the endless bush. Shapes moved through the long-abandoned walls as we sailed past, baboons the only dwellers now.

An elephant standing in the shade of a baobab tree reminded us that it was hot down there; ahead a lonely black lighthouse tower stood on a sandy hill. We banked steeply, gravity pushing us into our seats, and swept around the point where a splotch of green trees and pale houses appeared. A minaret rose among the roofs and white-limed walls. It was

From bones unearthed in the ruins, we know that cats lived among the first settlers a thousand years ago.

our first glimpse of the fishing village of Shela on the seaward tip of Lamu Island.

Three miles up a channel between two flat islands stood the larger town of Lamu proper, where a quay was busy with activity. Small boats moved in and out from a larger one at anchor. The engine slowed as we leveled off and approached a brown slash through the bush on the island across the channel. Wheels touched with a jolt and small stones rattled like bullets against the fuselage. We had landed at Manda, one island in the group known as the Lamu Archipelago, sixty miles south of the Somali border in the North Eastern District of Kenya.

A couple of hours later, my companion, Sieuwke, and I dove into the Indian Ocean, warmest of all seas. The tide was running out and we let it carry us toward the point. There we came ashore and walked along the ocean beach.

Soon the channel entrance and the village were out of sight behind the headland dune; the length of the island lay before us in an unbroken strand eight miles long.

Two sets of footprints, our sole human predecessors of the day, went out and came back again; the only other visible living things on the beach were a lone tern and the ghost crabs that scurried into holes ahead of us. We relished the feel of the sun and water against our skin; then we walked back along our footprints to have a look at the village of Shela.

Eighteenth-century ruins of what had once been grand stone houses stood between today's more modest thatch-roofed dwellings. These spacious stone homes must have been occupied by a people of substantial wealth. We crawled through a curtain of vines into one bat-tenanted ruin. Intricate plaster carving in geometric designs covered crumbling walls, and massive iron-hard wooden door frames were carved in graceful patterns, a tradition for which Lamu rivals Zanzibar in fame. Through one portal stood a cisternlike bath with an ancient Chinese porcelain bowl inset. The bowl was there to hold a last bit of water when the cistern was flushed out to be cleaned, thus saving the lives of the pet fish that were kept for mosquito control. This and the following I learned from the thin Lamu guidebook, *Lamu Town: A Guide*:

The most impressive part of these houses is their bathroom and lavatory system . . . Each house has at least two, many three, four, or even five. The lavatories emptied into deep pits, sterilized by the surrounding sand and sufficiently far removed from any nearby wells . . .

Washing water . . . was all taken away, either out through the streets in open stone-lined gutters, or down by means of underground tunnels which wound their way beneath other houses to the sea.

Each bathroom has a cistern, fed from outside through a conduit in the wall except where houses had their own interior wells (which many had). Water was stored in this cistern and the inhabitants stood beside it and dowsed themselves from it. . . .

These bathrooms were especially beautiful . . . The money spent on them can only have been equalled by the money spent on qiblas of mosques. Oddly enough, with their trifoliate arches, their chamfered pilasters, carved coral and intricately decorated niches, mosques are what the bathrooms often closely resemble. The link between cleanliness and godliness, always stronger in Islam than in Christianity, appears to have been particularly close in the Swahili world and is here symbolically illustrated.

It must be remembered that in the eighteenth century, bathrooms were a rarity in Europe, and domestic and civic plumbing and drainage arrangements of practically all western towns were squalid beyond belief. It is a sad but salutary reflection that eighteenth-century Lamu must have come very close to being that ideal settlement, nonpolluting and functioning with little or no disruption to the ecological balance, to which so many northern hemisphere settlements aspire and so few approximate even today.

As we walked through the old pathways of Shela, with its tiny population of less than a thousand people, its open spaces littered with plastic containers, rusty tin cans, and bottles, I was impressed by the impact of Western progress on an ancient civilization and by the irony of this trash as the most visible aspect of its modernization. Yet I could close my eyes to the rubbish and still see the beauty. Coco palms grew profusely throughout the village, and gardens of papayas, limes, mangos, and bananas were everywhere. Riots of multicolored bougainvillea cascaded over walls and roofs. Off the beach, twenty *machuas*—small lateen-rigged dhows— were moored, and other sails skimmed up and down the channel from Lamu town.

As we walked the path from Shela to Lamu town, a bright red bee-eater darted from the top of a tamarind tree, its long tail streaming, and grabbed a cicada out of the sky. Sickle-winged swifts bulleted through coco fronds and a kite hovered on motionless wings as if hung from the sun on an invisible string. We came into town from along the south seawall and found Lamu bustling with activity.

Boats jammed with colorfully dressed people were arriving from outlying islands. Passengers were handing parcels up to the quay along with children, goats, and

Carved wooden door frames were tradition for several hundred years. The oldest now standing date to the mid-1700s. The wood is iron hard, and even a cat can't scratch it.

Lamu town.

baskets full of fruit. Other dangerously overloaded small craft were casting off lines and sailing away.

The sand at the foot of the seawall was patrolled by stick-legged maribou storks, and we noticed an unusual number of cats checking out the flotsam. We remarked at their Egyptian look, but besieged as we were with so many new impressions, we let the observation pass for the moment.

The next thing to attract our attention was the activity around a number of huge stacks of timber. The shipment

Residential quarter, Lamu town.

of mangrove poles was a flourishing business when we arrived in Lamu (since then, mangrove export has been banned in the interest of forest conservation). The mangrove poles, called *boriti,* were stacked along the quay in blocks as high and wide as large buildings. Cut, limbed, and barked, the poles were of uniform six-inch diameter and ten-foot length, the dimension that determines the basis of traditional Arab house construction. The length has been the standard for hundreds of years, and *boriti* poles have been one of the chief return cargos—along with ivory, rhino horn, dried fish, coconut oil, and simsim—of the dhows that brought dried dates, figs, almonds, furniture, carpets, and carved chests to the African coast from Arabia.

The deserts to the north are devoid of timber, and mangrove wood is strong, hard, and resistant to the boring of insects. As a consequence of the ten-foot length (a practical span to stow aboard a trading dhow and a most available size), Arab houses are built in ten-foot modules. If a

This Bajun woman in her buibui is Muslim, as is at least half the population of Lamu.

room is to be wider than ten feet, the ends of the poles must be held up by pillars and cross beams, thus the rows of arches so characteristic of Arab or Moorish architecture.

A row of buildings lined the waterfront, some white-washed, others stained with the patina of time. One block back from the quay, we walked along the main commercial street where cafes and shops faced each other across a narrow lane. The way was crowded with people coming and going.

Behind us, I was vaguely aware of someone whistling. Pedestrian traffic began to dwindle as people stepped into the niches of recessed doorways. The whistling became more insistent, and a woman whose dark eyes looked out through a slit in her black *buibui* reached from an alcove and urgently touched Sieuwke's shoulder. Sieuwke looked around just in time and pulled me into a doorway. A donkey cart with a streetwide axle that nearly brushed both walls trundled by, and a wheel mashed a lime in the street. Whistling his warning as he ran behind, a boy's eyes caught mine as he passed. He glanced down at the squashed lime, telling me with his eyes that if I hadn't stepped aside when I did, my toes might have been just as flat.

The *buibui* woman's eyes twinkled with amusement, and in the friendly way of the Lamu people, she adopted us for the moment. If our alien appearance wasn't enough, our indifference to the cart had confirmed that we were strangers. The lady apologized for the near accident. The carts were a menace, she said, but they had the right of way.

The streets of Lamu had been laid out long before the invention of the automobile and therefore were far too narrow for a car. In fact, she told us, there were only two motor

Women of Lamu enjoy a social evening outside their harem homes.

Henna stain is used to decorate hands and feet for weddings and other festive occasions.

vehicles on the island. One was a Land Rover that belonged to the D.C. (the district commissioner, head of local government). He used it to drive along the quay—the only street wide enough—from his house on the sand hill south of town to his office near the jetty a couple hundred yards away. The Land Rover was the delight of Lamuites, its uppity inutility the perfect symbol of bureaucratic waste. The other motor vehicle on the island was a *pikipiki* (such a marvelously graphic Swahili word), a Japanese motorbike owned by the Department of Power and Lighting. Also the subject of public amusement, the department operated a diesel generating plant, which supplied (with breakdowns so regular as to be dependably undependable, as the lady put it) the island's electricity. On his *pikipiki,* the power company troubleshooter could, in theory at least, race from mishap to mishap to repair the failures of the moment.

That evening, before we hailed a taxi (in Lamu a sailing *machua* provides this function) to return to the small hotel in Shela where we had a room, I again turned to my Lamu guidebook. There I found some words that somehow captured the whole atmosphere in a single paragraph.

Shops are designed to catch the attention of the women when they go out after dark. (Except for those observing strictest purdah, almost all Lamu women are free to go out together between about 8:30 and 10:00 P.M. Conventionally only the older or less respectable are permitted out by day.) To walk along the Usita wa Mui (the street of shops) at night, when it is mainly lit by light streaming from shops packed with brightly coloured fabrics and other articles showily displayed, is in fact quite an experience, even for the most jaded westerner; it makes him realize that Lamu, for all its small size, really is another town like many he knows. Its magic is only enhanced by the knots of black-veiled women with limpid, laughing eyes (and peals of giggles) who make shopping expeditions their pretext to strike up an acquaintance with the world outside their harem homes, and who know a thousand and one ways to drop a veil if it suits them to do so.

Sieuwke and I were captivated by Lamu Island, and in spite of such shortcomings as irregular electrical service, telephones that rarely worked, and no automobiles, plus the impracticality of its remoteness, we led ourselves to the logically insupportable conclusion that rather than detract,

Occasional failures of public utilities in Shela are the norm. The telephone system, for example, could present problems to any repair person who lacks the perception of a spider.

these inconveniences added in their way to the charm. After all, if one wanted all the amenities of Los Angeles, then one could stay in Los Angeles. We thought it would be an adventure to have a hideaway home in a place such as this; the dynamics of mere daily existence would kindle one's energy and add a zest to life—or so we hoped.

That night as we lay in bed watching the Southern Cross winking in the dark sky and listening to the rustling of the coco fronds, I felt very far from my home in California. I remembered reading about another person—D. H. Lawrence—who displaced himself from his homeland to take up residence in a small Mexican village, a community culturally far removed from his own. I could recall the gist

better than the words, and it went something like this: No matter how long one was to live in this village or how sanctioned one was made to feel by one's neighbors—free to share in their problems, rejoice at their marriages and births, cry at their deaths—no matter how much one strived to become a part of the community, there would always remain a gulf. There would always be that feeling of being an outsider.

I went to sleep with the thought, wondering if it was prophetic, but knowing that in spite of its applicability here where the gulf between cultures seemed as wide as it could be, we were emotionally committed; we wanted to try.

The Prides of Shela

Lamu has changed little since its settlement. Old Lamu town, three miles up the channel from Shela village (where I concentrated my main cat study), has the look, smell, and feel, I imagine, of the place it was a thousand years ago.

Although the cats of Lamu have no real homes as such, many adopt a house where the human residents are tolerant and toss table scraps the cat's way. In our case, more than a few cats have shared our house over the years. Matjam (which means *tiger* in Malayan) came to us shortly after we settled in Shela. Katja (*little cat,* Dutch), Liza (because my lady fancied that the gray and white kitten looked as beautiful as Liza Minnelli in the film *Cabaret*), and Simba (which, of course, means *lion* in the local Swahili) were a few who stayed long enough to earn names. All these cats have been gone for a long time, the simple consequence of longevity. Visiting felines, of course, come and go as often as fish are cleaned on the cement bench outside the kitchen.

Simba stayed the longest. By his twilight years, old Simba had fought so many battles, suffered so many gashes, infections, split ears, torn lips, ripped eyelids, and all the other injuries that randy tomcats can inflict upon one another, that out of compassion we felt it was time to put him down. But the day of judgment was always put off until tomorrow.

In his youth, Simba had learned to open the latch to the refrigerator and help himself to whatever he fancied from inside, a clever bit of work that can put to rest any idea that cats are stupid or lacking in tactile skills. After years of suffering his singular aptitude, I put Simba to the supreme test by tying a lanyard to the refrigerator door handle and looping it over a hook on the back wall. This simple device was too much even for his talent, and I was justly proud of thwarting my feline rival for cold milk and leftover chicken. At times when the old cat sat straight-backed on the kitchen floor in that regal Egyptian pose, staring thoughtfully at the locked refrigerator, we wondered: Is Simba really trying to figure out the puzzle of the lanyard?

Then, early on a Christmas morning, when at noon our guests were due to arrive with hungry expectations of a turkey luncheon, such a rare choice fowl having been specially air-freighted from Nairobi, Simba finally solved the vexing problem of the lanyard.

As I came in from the garden with a handful of limes for the stuffing sauce and stepped into the kitchen, I saw that the refrigerator door hung agape. The turkey carcass lay on the concrete floor and Simba's tail protruded from the cavity beneath the breast, wherein he was noisily licking up the last of the liver.

He pulled out his head, stared at me for a triumphant moment, waddled into the garden, belched, lay down, and died.

We call her Liza because she's endowed with all the grace and beauty of Liza Minnelli in the film Cabaret.

It's often perceived that like cougars and tigers, house cats are loners. The domestic cat is also compared to the leopard—indifferent and remote. The males of all these wild species mark territories that they aggressively defend from other males of lesser social rank, and adults come together in friendly society only for mating.

But contrary to popular perceptions of standoffishness, the highly socialized cats of Lamu behave more like African lions than they do other felines. Like their distant larger rel-

atives, these feral cats live in groups that resemble exclusive clubs. Unlike lions, however, the cats do not share in nurturing cubs, nor do they cooperate with group hunting tactics; but otherwise, their gregarious life is much the same.

One such group of cats occupies the territory on the beach in the fishing village of Shela. The membership of this pride (if I may be excused for extending to my group of social cats a term generally ascribed exclusively to lions) varies in number from year to year but averages sixteen to

eighteen. I have spent many hours watching other prides in Lamu town, but the Shela group that I call the Mangrove Pride has become the focus of my study.

The sand of Mangrove Beach is lined by tidal rows of coral pebbles, empty sea shells, shards of Ming porcelain, and broken bits of old Dutch Leroux trade pottery. Two dozen small sailing dhows called *machuas* regularly anchor here, where they are stranded by low spring tides or ride high on their anchors when the water is in. A communal

The fishermen of Shela mingle and mend nets in communal huts along the beach. In front of one shed, a lone mangrove tree is the gathering point for the Mangrove Pride. In the shade of this tree at low tide, the cats—and usually a few fishermen—spend idle hours.

thatch-roofed shed occupies the beach above the high-tide line. Here the fishermen of Shela gather to mend nets, make sails, engage in boat-related carpentry, and socialize.

A row of plaster-walled houses fronts the beach.

Lamu cats depend upon fish and household scraps for food and do not hunt the island hinterlands far from their human providers.

Coconut palms, a couple of neem trees, a few flamboyant trees with their masses of bright red blossoms, and a row of casuarinas provide shade. Just in front of the fishermen's shed, standing tall with its roots in the water, a single mangrove tree is the focal point for the Mangrove Pride. It's in the shade of this tree at low water that the cats—and usually a few fishermen—spend idle hours.

I began my investigation fully intending to follow the mode of a scientific study, using the numerical system commonly used by biologists to identify my subjects. But I soon realized that so many cats in my study group had such dis-

Kinky is the only member of the pride to use the limbs of the mangrove tree as a lying-up place. It was the kink in her tail, not in her character, that suggested her name.

tinctive characteristics or personalities that obvious names practically shouted to be recognized, and I promptly let the numerical system go by the wayside.

In spite of a constant changing of the guard over the years—changes brought on by births, deaths, and diffusion—many cat characters stand out in my memory. At one time during my study five males and seven females were the hard-core members of the Mangrove Pride. The beach was their permanent home. The others were either kittens of the hard-core group or regular visitors—cats that sheltered and took some of their food in houses inside the village, but used the beach as their hangout.

There was Kinky, a female ginger with a kink near the tip of her tail and a conspicuous scar on her nose. It was the kink in her tail, not in her personality, that inspired the name. She was the only cat to have exclusive use of the limbs of the mangrove tree as a lying-up place.

Marmalade, an unimaginative name, if fitting for its perfect description, was all-over-orange with contrasting dark and light stripes, as in a jar of the wonderful tangy stuff. She was a hard-core female.

Lady Gray was pale rusty gray with blended russet patches. She was perhaps the most ladylike in appearance if not behavior, which I judged to be rather fast—even for a cat. She was, however, very beautiful.

Slim was a half-grown female ginger, sickly, and thin. She was the only cat in the pride who appeared to be in chronic poor health.

Bwana Mkubwa means *Big Man* in Swahili. (Kenya's first President, Jomo Kenyatta, was a Bwana Mkubwa, not

Lady Gray wears silver slippers on her toes.

Somehow, without suffering the usual battle scars or earning combat credentials, Midnight ultimately achieved top place in the Mangrove hierarchy.

Ink Spot takes the classic stance when confronting a dog. One rarely sees this intimidating attitude in cat-to-cat encounters; it's saved for threatening predators.

because he was big in size but in importance.) This cat was a large male both in size and heart. An elder of the tribe, he wore a large black spot like a medal of valor (or so I fancied) on his white shirt.

F-6 was a white female with tiger-stripes on gray tabby patches. Her name is a hangover from her original numerical designation. Because of her infrequent appearances on the beach and because I originally misjudged her to be an undistinguished cat, I let her naming under the new system slide. Maybe if she'd been around more, my opinion of her as an uninteresting character would have been different from the beginning. More about her dauntless character later.

The distinctive markings of Son of Ink Spot leave little doubt that the genealogical tree has sprouted a new branch.

Midnight was a jet black tomcat who—somewhat surprisingly considering his apparent youngish age—was second in male dominance, beneath only Bwana Mkubwa. He wore none of the facial scars one would expect of a high-ranking male.

Big Tom was a large, older gray-and-white tom with enough scars to show heroic battle credentials. It was difficult to determine his precise rank, but it was not tops in the pecking order. He was easily distinguished by a permanent limp from an old injury to his left front paw. Assuming a hereditary linkage because of the similar black side spot on white background, he may have been a son of Bwana Mkubwa.

Ink Spot, another character with a distinctive black blotch on his side, was probably a son of Big Tom. Because of his large size and general bearing, I originally assumed Ink Spot was number one in the male pecking order, but he deferred to the smaller and younger Midnight. From Ink Spot, I learned a lesson: Never make off-the-cuff assumptions about a cat's social ranking. One must observe their interactions to know the truth.

Son of Ink Spot was the fourth tom to show striking similarity of markings to his probable great-great-grandfather Bwana Mkubwa. He wore the same insular black spot on his flank, a characteristic that suggested a continuity of at least four generations in the pride.

Tortoiseshell was a female of considerable beauty. An infrequent visitor, she was not, unfortunately, one of the hardcore. (I say unfortunately because I could easily have fallen in love with Tortoiseshell, my favorite cat color, and I'd like to have seen more of her.)

Shoulders was a hard-core young male with dark gray shoulders outlined as if painted on white paper with a sharp-tipped brush. A bit more than half the size of Ink Spot or Big Tom, he was a youngster just beginning to assert his mettle.

F-13 was another hangover from my original identification system. A hard-core ringer for Kinky, except she had a straight tail and no nose scar. Never showed me enough character to rate a proper name.

Bibi (the Swahili label for *wife*) was the black-and-white mother of three unweaned (at time of writing) kittens. I first called her Skunk Face because of the resemblance (black face with a narrow white stripe from nose to forehead), but she was too amiable for the name to stick.

Shoulders.

Bibi's Kooky Tabby Kitten, usually referred to simply as Kooky, was the young prince that I have high hopes will someday be the king of the pride.

Bibi's Spooky Tabby Kitten or Spooky was a female that I could have called Docile or Indifferent, but Spooky stuck.

Bibi's Tortoiseshell Kitten or Safi (meaning beautiful) was the third of Bibi's kittens, intermediate in precociousness between her tabby brother and sister. This aloof little princess could be wearing glass slippers. My hopes for Safi were that she would pass on the genes of her coloration to many offspring. In this respect alone, Safi grew up to disappoint me. Her first litter of four kittens produced only one like her in color.

Ginger was lighter than Marmalade, more all-over-orange without darker stripes. She was a hard-core female cat of little character or distinction.

Mystery was a large male ginger with a white tummy who visited the beach regularly, but usually at night and customarily (but not always) with a furtive attitude, as if he expected to be jumped at any moment by a dominating beachmaster. His daytime hangout was with the Dutchman's Pride.

Long John Silver was actually a high-ranking member of adjacent Badi's Pride, but he visited Mangrove Beach and interacted with its members so frequently that he rates inclusion in this register of pride members.

I always thought of cats as sleeping through the day and wandering through dark alleys at night in search of rats and mice, yowling at the moon in frustrated nocturnal erotic encounters, and carousing in all sorts of ways. If such night-time debauchery is true in most cases, then the Mangrove Pride differs widely from other groups of feral cats. Lamu cats have adopted a way of life in tune with the activities of the people they depend upon for food. Shela fishermen are not nocturnal, and neither are the cats of Shela. Lamu cats are more in tune with the changing of the tides than with the setting of the sun.

Bibi, meaning wife, *and her brood. Having raised at least five litters, she has the experience to exert strict discipline. At the same time, Bibi is not a nervous mother, and her kittens are allowed a lot of room for experiment and play. Her progeny are known for character and individuality.*

Each day, I began my studies at dawn when the cats were scattered up and down the beach, sleeping in the shade of beached *machuas* or on stones under the mangrove tree if the tide was low.

If the dawn tide was high, the cats would be asleep under the fishermen's thatched roof or on the steps of the mosque—anywhere they could keep the beach with its promise of fresh fish in sight and still keep dry.

I soon learned that low tide was when something was most likely to happen. Not yet formed into groups, the cats would sleep until the tide went out, at which time either one or both of two things of interest to the cats might occur.

Most likely, somebody would come down to the fish-trap and begin to remove the high tide's catch. That was good reason for a general movement of all the cats to the place on the beach where the fishtrap catch was scaled, gutted, and sold off. This operation was always carried out at exactly the same place—inside the hook of the fishtrap at

Kooky, the extrovert, on the left and his brother, Spooky, the cowardly lion. Their sister, Safi, is a cool tortoiseshell with all the potential to become a royal queen.

Long John Silver, alpha male, a warrior who proudly wears his battle scars. Badi's Pride, a rough crew of pirates including Blind Pew, Blackbeard, and Captain Hook, occupies the territory inland from the Mangrove Pride.

The Mangrove Pride breaks many familiar presumptions about cats. Their movements are coupled to those of their providers, the fishermen—they sleep at night and awake at dawn.

Watching a fisherman's sail coming in: Lady Gray, Bibi, Kinky, Shoulders, Midnight, Big Tom, Ink Spot, and Ginger, hard-core cats of the eighteen-member study group.

the high-tide line. The cats never moved rapidly to the fish-trap assembly point, but they always moved en masse from various points, so their approach to the fishtrap was like a slow wave sweeping along the beach. All the cats walked, never hurried, and they positively never ran. They were far too suave for that.

If, however, the fishtrap overseer didn't show up, there were always the boat fishermen.

Whether the fishtrap man made his appearance or not, the landing on the beach of any *machua* started a movement of all the cats similar to the fishtrap parade. Again, the most striking thing about these mass movements along the beach was their *casualness,* as if no beach cat seemed to want to appear to have been hustled. Ink Spot was the coolest. He strode slowly down the beach with regal dignity. Marmalade was the uncoolest. She had a certain polish in her step, but sometimes I detected a quickening of her stride, as if she was trying to be first in line. She has even been seen to overtake, something simply *not done* in the best of Lamu cat society.

If Marmalade was uncool, Kooky was simply gauche—but he was still a kitten, after all. As often as not, he ran the opposite direction to the general drift of his elders. They paid him not the slightest glance.

*The catch is cleaned at the base of the fishtrap—fillets
to the buyers, guts and tails to the cats.*

Some days the fishtrap man does not appear, but the landing on the beach of any machua *creates a movement similar to the fishtrap parade. Again, the most striking thing about these mass marches along the beach is their decorum. No cat seems to want to appear to be hustled.*

III The Brotherhood

For all their daytime observability, when darkness fell, the movements of my cats became a mystery. Many of the shadowy places they visited at night were closed off to my eyes, and I could only catch glimpses and gather vocal hints at their nocturnal activities.

Sometimes in the morning, it was clear that there had been business—bleeding facial wounds were evidence that wails and growls during the night had been the voices of a fight. But the midnight sounds of cat confrontations only hinted at what was going on and made me want to know more.

I had at first assumed that these cats wandered widely throughout their available habitat. A local resident told me

Ordinarily males fight with other males and females fight with females; but here an incongruity as Ink Spot threatens the female F-6.

that the mangrove cats spent their days on the cool beach and roamed through the village to the dump behind the village to feed at night. When he took trash from the kitchen out to the dump in the evening, cats were usually there. He assumed they were the same ones—and at first, following his reasonable deduction, so did I.

But they weren't the same cats. I never saw a beach cat at the dump and I never saw a dump cat at the beach. In fact, I never observed a cat's range to extend more than a hundred yards. Although difficulty of observation may account for this in part, I've learned that Lamu's feral cats occupy very small ranges, limited to the houses and gardens—and in the case of the Mangrove Pride, to the beach—where they get their food. The most far-ranging cat I know is Lady Gray of the Mangrove Pride. I've seen her cross our garden many times—that's already some eighty yards from her beach hangout—and when she crosses, she's clearly headed for somewhere farther along on the other side. But I've been unable to discover her destination or if she truly has one.

One curious sidelight to the travels of Lady Gray is that she bears no scars. I now know that not only male Lamu cats fight each other over territory; females can be equally fierce in defense of hallowed ground, as proven both by firsthand observation and by the facial wounds of female cats. How is it that one of my most widely traveled cats bears no marks from territorial encounters? I can't say for sure, but perhaps it has to do with demeanor. Not only is Lady Gray one of the prettiest cats in the group, but she knows it. As confident as a beautiful woman at a cocktail party, she carries herself along the dark walls of Shela's dangerous no-man's-land with cool assurance, aloof, secure in the knowledge that whatever is out there, she'll be okay. Is it her attitude of composure that gets her by?

There are several conceivable reasons for cats to wander: for food, for play, for sex, and for those low on the social ladder in one territory to climb higher somewhere else. Within the Mangrove Pride, food and play could be taken care of within the group. Sex and social climbing really go together, at least for the male members. The older dominant males are the ones who usually get the sex, so young males must strike out on their own in search of new territories, hold their own with or beat the older males in physical confrontation, or win a place by cunning. Only a small percentage of young males have a chance of achieving a dominant sexual status, as the harem system—one man, several wives—prevails as much for the cats outside the walls of Lamu as it does for the humans within those walls. (Both Muslim law and the customs of Lamu animists allow several wives).

One way of eliminating male competition at an early age is for tomcats to kill kittens. If given the chance, they will sometimes kill the offspring of their own mates. It has been shown that this genocide is often sex selective—that male kittens are the principal victims of their fathers. I saw only one serious confrontation between male and female cats that could have been associated with this behavior. The fight was between Ink Spot and F-6. During the preliminary face-off, when F-6 rolled onto her back, I noted that her teats were conspicuously swollen. Now I knew the reason for her fierce aggression—she had young kittens hidden away somewhere. F-6 got soundly thrashed by Ink Spot in the battle, but the confrontation served its purpose well. She distracted the predatory male and saved her kittens. And to think I had believed her to be so much without character I hadn't even given her a name, when indeed she was the bravest of cats.

An interesting sidelight to this cat fight was the reaction of the fishermen. The battle took place on the beach directly in front of the fishermen's hut, and several men were sitting inside it at the time. By now, through my more-or-less constant attendance on the beach, I had at last become an accepted part of the landscape. But my study of the cats and frequent picture-taking was still the best amusement for the locals since the last time a visiting yacht had grounded on the reef. Although cat fights weren't all that novel, *I was,* and all eyes were upon me.

When I arrived on the beach and heard the wailing of the two cats in confrontation, I hurried close with my camera and began shooting. After two or three minutes of face-

Only an hour after the battle, the same two cats sat down whisker-to-whisker and shared a fish head as if nothing had happened.

to-face threats, vocalizations, and paw jabs, Ink Spot rushed F-6. F-6 ran but was quickly overtaken, and as they rolled on the sand, fur literally flew. Then they broke off and faced each other in nose-to-nose standoff, growling and hissing.

I moved quickly to a better camera position, propped my back against a *machua* to steady my lens, and resumed snapping. Then F-6 made a break for safety and Ink Spot was after her again. Surprisingly, they ran straight in my direction, and as they rushed toward me I had to move quickly or be caught in the middle of a cat fight. The sight of me

When sociable feelings influence the pride, personal space shrinks, but still nobody steps on anyone's toes.

scrambling to keep a flurry of claws out of my lap was cause for a round of laughter from the fishermen's shed.

Then, as the two cats dashed between my feet and disappeared around the bow of the *machua*, one of the fishermen ran down with a coil of rope, whipping it against the sand, and shouting at Ink Spot. As adults will interfere in

At other times, privacy is graciously granted.

the brawls of children to whom they have no relationship, the fisherman felt compelled to break up the cat fight, a bit of human-cat interaction that was as illuminating to me as the fight itself.

The hours I spent watching the cats gave me many answers, but I still had questions about some of the behaviors I had seen. Male cats, for example, aren't known for beating up females, so why had Ink Spot attacked F-6 with such ferocity? When challenged over the issue of the kit- tens, couldn't he simply have backed off and gone his own way? But no; he had to show his machismo and give her a thrashing. Only an hour after the battle, I had seen the same cats sitting side by side as if nothing had happened. How could one explain such behavior?

I turned to the work of animal behaviorists who had studied cats extensively. After depleting a stack of publications in which I had found the going rather hard, I opened at last a scholarly treatise on cats with language that embraced some wonderful everyday usage.

I didn't find exactly the explanation I was looking for within the covers of Paul Leyhausen's book *Cat Behavior* (as Ink Spot's conduct seems to have been exceptional), but I did discover a lot of scientific interpretation of what I had seen along with a whole new world of cat lore.

The first of Leyhausen's remarks that caught my eye were so in tune with my own feelings that I knew right away I had found a friend: "The following domestic cat observations . . . lead me to assume that in the matter of social intercourse what is sometimes true is not always true."

Describing the behavior of an experimental cat that had been placed in a strange room with another cat already familiar with the surroundings—the newcomer thus being at a disadvantage and therefore initially submissive—the author says that the stranger invariably "exhibits a remarkable form of behavior which I call 'looking around.'" After browsing the scientific literature, I'd come to expect stilted language. To define such behavior as Leyhausen had seen, most scientists would use stuffy jargon such as *a sex-specific random ocular avoidance mechanism*. To find a technical description so engagingly put as "looking around" was a refreshing surprise.

The scientific meaning of "looking around" in this case was made clear by the following account: "Moving its head with the alert and contented expression with which . . . a well-fed cat sits at a window surveying traffic, it looks around everywhere in the room except in the direction of the resident cat, as if to emphasize how harmless it is."

In this stranger-in-the-established-community experiment (a classic textbook situation for novelists, too), Leyhausen goes on to describe a typical encounter by saying that the highest attainable place in the room is the most advantageous. If at the beginning of the encounter the strange cat can get itself up onto a chair, for example, while the resident cat is on the floor, the newcomer immediately achieves a degree of dominance.

With its former superior position of territorial ownership thus usurped, the resident cat is intimidated. At this point of near psychological balance, the cat with the strongest personality can rule.

"'Looking around' is obviously connected with the dislike every cat has of being looked at directly," Leyhausen says. "If a cat is being observed secretly and it suddenly becomes aware of the observer's gaze, it is quite certain to stop whatever it was doing and will recontinue only in a noticeably inhibited and hesitant manner. If a cat is stalking prey [and] suddenly realizes that it has been detected, it will immediately straighten up, 'look around,' and feign indifference."

I had seen this behavior occur several times with outsiders who had blundered into the territory of the Mangrove Pride—and once when I had followed Ink Spot on a ramble through the village. He rounded a corner and unexpectedly came face to face with the dirty-colored tom

I called Ugly. If not on unfamiliar ground, Ink Spot was out of his own territory, and there is simply no better word or phrase to describe his reaction. He sat down as if Ugly didn't exist and "looked around" for about four minutes, never meeting the eyes of the resident tom. Then he got up casually and strolled away, angling off in a different direction than his originally intended path, which would have taken him closer to the resident stranger. In this way an unpleasant confrontation was avoided.

If Leyhausen could turn this rather complicated social intercourse into a concept so simple, I wondered, what could he tell me about the nocturnal habits of the Mangrove Pride that had been so difficult for me to see? Here again, I was struck with his clarity. He had observed with scientific accuracy some of the same behaviors I had seen and had drawn the same conclusions.

A common behavior in groups of feral cats all over the world is called "looking around" by the biologist who first reported it, a delightfully unscientific-sounding phrase that couldn't be more technically explicit.

Members of the Mangrove Pride "sitting around"—a gathering of passive social exchange that has been observed in other groups of feral cats around the world.

At nightfall there is often something which I can only describe as a social gathering. Males and females come to a meeting place adjacent to or situated within the fringe of their territories and just sit around . . . There is very little sound, the faces are friendly and only occasionally an ear flattens or a small hiss or growl is heard when one animal closes in too much on a shy member of the gathering. Apart from that there is certainly no general hostility, no threat displays can be seen except perhaps when a tom parades a little just for fun.

Leyhausen had no doubts that these were friendly, sociable get-togethers. But, in the same way I puzzled over my cats' inconsistency of behavior, he wondered at the enigma that "members of these same populations could at other times be seen chasing each other wildly or even fighting."

He noted that the same "urge for social 'togetherness' exists also in those wild species [such as leopards and tigers] in which, according to all available observations, mutual repulsion is much stronger than it is in domestic cats."

The cats' evening social gatherings are generally in the nature of a tea party or kaffeeklatsch, and males, once the social ranking has been established, appear to tolerate trespassers even more agreeably than the ladies do. "Fierce

Ginger is the first to break up a session of "sitting around" and head off on her own.

defense of the home and the home area is usually exhibited . . . in domestic cats only by females rearing a litter."

In describing the behavior of tomcats, Leyhausen says that serious fights are likely to take place only at the time of a first meeting, when the relative strength of each cat is sorted out. Thereafter, once each tom's position has been established, arguments are settled by display and not fierce fighting.

In this way, after some initial combat, those toms who aren't beaten down to become complete outcasts (pariahs, as Leyhausen calls them) form a kind of order, or brotherhood:

They gather in friendly convention . . . and even in the mating season seldom fight to the bitter end.

The picture is strikingly different if, within the established neighborhood, there is a young tom just crossing the line from adolescence to maturity. The established tomcats of the vicinity, alone or in groups of two or three, will come to his home and yell their challenge to him to come out and join the brotherhood, but first [he must] go through the ini-

tiation rites. The challenge is not the piercing, up-and-down caterwauling of the threat display but rather softer and seems to have a good deal of purr in it, as if it were not merely challenging but also coaxing. In fact, the sound is hardly discernible from the call by which a tom tries to entice a female in heat to meet him. If the youngster lets himself be persuaded, hard and prolonged fighting ensues. This is in fact the situation in which most really bitter fights occur. And since the novice, who feels his strength growing from day to day, will not accept defeat as any sensible adult would, he will at first be beaten up and often badly injured. But the wounds have hardly closed before he hurries to battle again. After a year or so, if he survives and is not beaten into total submission, he will have won his place within the order and the respect of his brethren and now sets out in his own turn to 'teach the young heroes a lesson.'

Although each male would prefer to be a lone king of the pride, the fights that would be needed to achieve such ascendancy are simply too risky. The lower ranking defer to the more powerful often enough that the alpha male feels no need to assert his authority with all-out combat. In turn, even a low-ranking male can occasionally cop a copulation, so that he does not feel impelled to risk mortal combat to depose the leader. Thus the brotherhood is nothing mystical, "but rests on a very real balance of power, risks and deterrents. It can be formed only if there are several males of almost equal strength, so that victories and defeats are decided by a narrow margin and it might cost a higher-ranking male his superiority if he provoked an inferior to the point of actually making him fight."

All of this behavior I had noted in the Mangrove Pride. But until I read the scientific evaluation, it hadn't always dawned on me exactly what I had seen. And I very much fell for the concept of an animal brotherhood. Here, without any note of anthropomorphism, I had stumbled onto a scientific work utilizing poetic terminology with a humanizing flair. "Looking around" and a brotherhood of cats—how wonderful!

One of the trade-offs in exchange for the inconveniences of life in Africa is the luxury of servants. In America it almost seems one should apologize for the indulgence of a live-in housekeeper, but servants are necessary to life in Africa where a house without staff is considered a great oddity.

When our house at Shela was completed, our next concern was to employ a good servant. Sieuwke tried a couple of applicants but was unable to find a good match. Then a small young man who said his name was Sharif Konde knocked at the gate and asked for employment. Sieuwke could not help but notice that his left arm ended just below the elbow. Obviously, a one-armed housekeeper was not the most practical choice.

But there was something about Sharif that caused Sieuwke to consider his application seriously. He told her that he was eighteen years old, came from a small farm a hundred miles to the south, that he was of the Giriama tribe, and he apologized for his inability to present references. The reason was simple—he had never held a job before. In spite of his overwhelming lack of qualifications, Sieuwke hired him. "He has a nice face," she said. It was an insight which was not to let her down.

With patience that I found incredible, an aspect of Sieuwke's nature I had not until now fully appreciated, I watched her teach young Sharif the strange facets of European etiquette as to how to serve at table: the mystifying customs of presenting plates from the left, collecting from the right, the curious importance of placing knife and spoons on one side of a place setting, forks on the other.

What could this have meant to Sharif, who had been brought up in a mud hut in the bush? Surely our customs were as strange and exotic to him as the tradition of sitting on the earth and eating with the fingers from a common bowl would seem to us.

He learned to make beds, launder clothes, polish floors, and clean house. But cookery seemed not to interest him in the least.

After Sharif had been with us for half a year, he observed one hot evening as Sieuwke sweltered in the kitchen that the house should have a cook. When Sieuwke replied that she couldn't agree more, Sharif was sure the time had come to offer a solution. Through his infectious grin he told her that his younger brother, David, had been in Lamu for a month working as a mason's helper. Although David was not trained, he would like inside work. Perhaps the Memsahib could teach David to cook just as she had taught Sharif to be such a perfect housekeeper?

Not one to heap praise where it's not entirely due, Sieuwke replied, "Well, hardly *perfect,* Sharif. Not yet. Remember last night you snatched away the soup again before the guests had finished."

"But madame, you yourself said that if any soup was left over, it should be fed to the cats. If I hadn't taken away the bowls when I did, the cats would have gone hungry."

Sieuwke shrugged hopelessly. "Okay, Sharif, I'll talk to your brother."

David was the physical opposite of his brother. Sharif was short and slight; David was tall and muscular. He seemed made for the mason's work he was doing rather than for

Sieuwke hired Sharif on the merit of a single qualification: "He has a nice face." It was an insight that was not to let her down.

the inside work he preferred. But he said his mother had taught him to cook Swahili food and that he was sure the kitchen would suit him fine. Sieuwke said she'd give him a month's trial.

The first traditional Swahili dishes David prepared for us were a delight: fish baked with slices of onion, tomato, and lime, with sauces of coconut milk and strange spices I didn't know—turmeric, cumin, cardamom, coriander, marjoram, ginger, and peanut, all laced with a subtle curry. David was a born chef. "His fingers seem made for it," Sieuwke said, and in a short time she turned the kitchen over to him.

When the time came for us to leave Lamu for our annual trip back to America, we worried about leaving David and Sharif to look after the house on their own. Some of our neighbor's staff grew bored and got sloppy when left without other people in the house. But there was always work to do if one was diligent: cupboards to be swept of spiders, bedding and mattresses to be aired to keep down the mildew so prevalent in the humid climate, furniture and floors to be mopped with kerosene to discourage boring insects. There was the garden to be watered, trees and shrubs to be trimmed, cats to be fed, and a hundred other things. Could we depend upon David and Sharif to fulfil their newly learned obligations? All qualms were put to rest when letters from Lamu began to arrive in California.

David, who had attended school, writes and speaks English. It is David who keeps the household accounts, balancing shillings paid for cat food, soap, and polish with the income from limes and papayas sold from the garden.

As years went by, David's letters took on a familial charm. We looked forward to their arrival with as much pleasure in their style as content.

There is a whole family's saga in the years represented by this correspondence, and David's letters tell much about the African people. His writing describes the universality of the human experience, the worries about finances, courtship, marriage, childbirth, illness, death. Yet the problems of the family Konde are so very different in the details, revealing the vast separateness of our cultures.

Nyumba nyuma ya misiquiti, the house behind the mosque, as the villagers call our home, has always been known from the first day of occupancy as the Bislet house because Sieuwke, whose surname is Bisleti, opened the residence during a time of my absence. Why the ending vowel of Bisleti has been dropped by our Swahili neighbors is a mystery, as the reverse practice is the usual rule. English words converted to Swahili usually pick up an ending i— *soksi* for socks; *futi,* foot; *chisi,* cheese; *daktari,* doctor. In Sieuwke's case, they had the *i* going in—Bisleti—and chose to drop it. Memsahib Bislet she will forever be called in Lamu.

David addresses all of his letters to Sieuwke.

Dear Memsahib Bislet,

It is with great pleasure to have this fruitful opportunity to let you read from me once more.

Thank you for your letter dated 25th Feb. which reached my palms on 1st March as I was expecting one. Before I can busy my pen, I hereby give briefly about my living and activity as well.

I am in health and my duty attendance is punctual. But of all that, most of my worries are based on your daily living and activity in both as we are living far districts apart from each. But nevertheless, I hope you and Bwana Jack are very well.

But something to amaze you, from 19th to 24th, Simba had completely become a nomadic cat. I remember to have missed Simba in my sight for five days. But since I was concerned, I reported to Memsahib Kay on 21st and was told to keep on searching for it. Because I never knew where Simba had gone, I also became nomad and looked for him, this way and that did I go in the whole village of Shela, but never did Simba come into my human sight.

Excuse, please may you turn over and proceed.

Reaching far from the previous page, very early in the morning of 25th Feb, I was struck dumb and rooted to the kitchen room when I was making milk for cats. Turning around to the kitchen table I saw Simba sitting on the table as though its were bird on a tree looking for some ripe berries.

How happy I was for having Simba into my sight again? Keeping apart with the above already written, I wish to turn and express much about the plants.

In watering plants I am struggling hard to achieve a credit for my industrious work. In the case of my being alone,

As a young tom, Simba had no enduring affiliations among the wild prides—ties to Liza, Katja, and Matjam were enough to keep him from gallivanting. But in his golden years, Simba sometimes yearned for greener pastures and hit the open road.

Sharif being on leave, I beg you to have no doubt, because for your plants watering is daily done.

Backing again to the house I am working in, all the bed sheets were folded neatly on 2nd of the present month. According to my imaginary timetable for the housework, I have been polishing the house floor and furnitures once in a week since you left me. Nevertheless, the drying or airing of the mattresses and pillows is done by me once in a week but in different days. Dusting and sweeping is daily performed. The checking of dudus [insects] is widely proved. In the early days after your departure, we used to seeing many kinds of dudus, but because "Johnson Hit" is often used we find none of the insects. Moreover, the metals are usually applied with "Brasso."

I am about to finish painting the house and plentiful glad because it is looking loveable white. The house is so much white, such that a stranger should say, it is a new built house. Don't you think it's a nice idea to have the main bedroom painted too, as it does need?

It is definitely my first time to paint a house and according to my proof, fundis [professional craftsmen] are no better than the miracle I am doing.

Already, about two or three persons asked me if I could do such a paint for their houses, but my answer is, "This is done specially for the long-absence and loving employer."

I beg excuse for re-writing this sort of letter, being its reason, reminding you the date for me to go home for a well and firm-fixed promise with my parents is no less than two weeks to reach. In their reply, my male parent has said that in case I fail to go home on either 4th or 5th of the next month, he will abandon to call me his second born, but an idol representing the lost ancestors of the home due to my failure to promise.

Besides my father's advice, my mother who is madkeen to get a second daughter-in-law, said that would give thanks to my employer if would let go for that last offer, otherwise I remain single till my 30s.

Before we put pen down to rest, we or I hereby inform you that the pili-pili plant died and a stone is on the spot. Whereas the jasmine plant has not died yet and was planted out side on 5th March, under care according to your instructions.

Your faithful servants,
Young David and Elder Sharif
Prepared and jotted by David

Beneath the Mangrove Tree

I stood near the mangrove tree looking down at Ink Spot lying asleep on a rock in the shade. I strolled closer and stood over him, surveying the beach and making notes on who of the Mangrove Pride was present. Ink Spot was used to human feet being near him, but even in sleep he probably sensed that I was standing no more than a few inches away. Every day of his life, dozens of feet had walked close past him at his various resting places. And they weren't feet that ever did him anything mean, like take a kick or tread on his tail.

Ink Spot had always seemed so aloof. At the time I assumed he was next in the line of hierarchy to take the throne when old Bwana Mkubwa passed on. He seemed so noble in his demeanor, a young prince fit and ready to achieve the monarchy.

I couldn't imagine haughty-looking Ink Spot as ever having been touched by a human hand. Not that he was *above it,* only that he seemed too, well, untouchable— there's no better way of putting it. Despite my instincts and better judgment, on an impulse I bent down over the sleeping cat and put the tips of my fingers between his ears,

Haughty-looking Ink Spot. Despite his aloof demeanor, he's one of the few pride members to allow the indignity of human touch.

ready—expecting, actually—to quickly pull back my hand when he snapped at me or rolled over and took a swipe with his claws.

Instead, Ink Spot barely opened his eyes, looked at me through the slits, and didn't move. I rubbed between his ears; he settled deeper onto his stone, and I got down to some serious stroking. If Ink Spot wasn't purring, everything in his body language was talking to me with a purr.

I was surprised by Ink Spot's acceptance of the human touch. I had always considered all of the Mangrove Pride to be on the wild side, and I had never before tried to touch one. Later experimentation proved that of the eighteen cats in the pride, only two, Bibi and Ink Spot, really liked to be stroked. A couple of others would allow a brief touch and then pull back, and the rest behaved as one would expect, as the semiwild animals they are. I wondered what had happened in the prior lives of these two cats to make them so different?

The tide was running out, some fifty feet of hard-packed flats already lay exposed, and the fiddler crabs were hard at work clearing their holes of the sand that had been washed in by the high water. They scurried from their penny-sized dens with armloads of sand, dumped it a couple of feet away, and returned, back and forth as if each one was tethered to its hole by an invisible yoyo string.

Down the beach, I noted Bibi's three kittens. They were now more than a month old, and as usual, Kooky was separate from his two less-liberated kin. He was far out on the tidal flat chasing crabs. Safi and Spooky watched from the security of their den—an abandoned overturned boat.

I had always wondered if in hard times, during weeks of prolonged stormy weather or when strong southeast trade winds blew, when the sea was too rough for the *machuas* to go out for fish, if the cats preyed on this abundant source of food? And what about the many small copepods and other crustaceans that lived among the rocks and jetsam at the high-tide line? Just what *did* these cats eat aside from fish?

To Ink Spot's apparent disappointment, I stopped stroking him, stood up, and moved closer to watch the kittens.

Kooky was having a ball. When one fiddler dove into its hole, another crab immediately appeared from a different burrow, and he chased it back. And with that toy gone, another was quick to appear—a game without end.

The inch-and-a-half-long fiddler crabs were no doubt named by an imaginative observer who noted the similarity between the single large claw of the male crab—one end of which, when the crab is at rest, lies tucked under his chin—and a violin. In times of procreative repose, the tiny other claw is the one that does all the work, moving rapidly back and forth between sand and mouth picking up bits of food. But during the mating season, the big claw comes into play. At this time, each male fiddler spends a good part of the day waving his "fiddle" high in the air to attract the attention of one of the many parading small-clawed females. Each female wanders rather aimlessly through the colony, her seductive stroll seeming to encourage the males near her to increase the energy of their gestures. As she approaches, they burst into a crescendo, so that she is constantly encircled by a huddle of madly fiddling males. There is no way for a human to see what it is in the males' provocative gesture that causes her to make a choice, but eventually she accepts the invitation of a male and pauses, facing him. Now the claw waving of the encircling spurned suitors becomes highly agitated, growing on frantic. But she gives them the cold shoulder and follows the chosen one into his hole. Nothing coy about this dialogue. What goes on beneath the sand is anyone's guess, but I assume that all the scurrying back and forth cleaning out burrows is to make a comfy boudoir.

But it isn't only to entice the females that the male fiddler crab is armed with a large claw. He uses it to spar with other male fiddler crabs and, although he would always prefer a safe retreat when approached by predators, he also uses it to defend himself in a pinch.

Kooky was quick—too quick for his own good. He charged between a crab and its hole and cut off a large and aggressive fiddler. The crab feinted right and left, running sideways, executing some impressive high-speed footwork, but the cat was quick and clever, shot out a paw, and rolled him over.

The cat danced left, jabbing his paw, and flipped the crab back onto its feet. Kooky rolled it over again, back and forth a couple of times. He stood up on his hind feet, pawed the air like a sparring boxer, and pounced. Then the fiddler's big claw closed on Kooky's paw. The kitten tried to shake it off; it clung. And the sharp pinch must have hurt plenty. Kooky tried to bite it off; his second mistake. Now the fiddler was stuck to his nose, and with a crab attached to that

Hors d'oeuvre de longhorn beetle. Mice and beetles occasionally supplement a fish diet. Geckos make good playthings, and when caught they shed their tails, which wiggle excitingly. This crunchy caudal part is also good to eat, explaining the many tailless geckos chasing moths under the lights at night.

sensitive organ all the fun was over. Kooky rolled onto his back, both paws scratching madly. Intense vocalization, as the ethologists would say, was noted by the observer.

When at last the crab let go and scurried back into its hole, Kooky raced up the beach and disappeared into the shadows, rejoining his siblings under the overturned boat. His two more passive kin had seen it all, and I wondered if cats can learn by example, something that my observations never revealed. If they can, the two student kittens must have gotten an unforgettable impression about how to treat fiddler crabs.

And if they didn't learn it by observation, I wondered if these firsthand lessons with the most convenient, most irresistible playthings on the beach weren't the reason for all Lamu cats' complete lack of interest in crabs for the rest of their lives.

A few weeks later, I watched Kooky chase a low-flying cicada around and wondered where these cats draw the line when it comes to playing with small living things. If cicadas are fair game, and crabs aren't, what else is or isn't? An easy experiment proved that the kittens had already learned that

Donkeys, marabou storks, goats, chickens, ducks, rats, and cats share the trash piles, occasionally all at the same time.

cockroaches are no fun, whereas long-horned beetles, in spite of their large size and intimidating mandibles, are. And they're tasty, too. Geckos make good playthings, and they shed their tails, which wiggle reflexively when detached—almost as much fun as their hosts. The crunchy caudal part is also good to eat. No doubt this accounts for the high ratio of tailless geckos we see chasing moths under the wall lights of Shela at night.

In my search for answers about Lamu cats' feeding habits, a survey of the literature turned up some interesting facts about cats and rats that throw considerable confusion into everyday views on the subject. Insofar as their rodent-catching propensities go, cats are mousers, not ratters. Several studies came to the same conclusion regarding cats as controls on rat populations. Ordinarily, a cat will not take on a full-grown Norway rat (the most common variety, also known as brown rat, wharf rat, or house rat). A big rat will fight back and make the ordinary cat turn tail. Cats will, however, keep a rat population under control by killing young rats if existing populations of adults are eliminated by other means.

In Lamu town, where there are several dumping areas with accumulated piles of trash, I have seen cats, rats (large, fully adult specimens), donkeys, chickens,

Newly hatched chicks are safe among hungry cats because of the ferocity of the broody hens. The chicks are unmolested if they stick close to mother.

marabou storks, and goats all picking through the trash at the same time. Even more surprising than cats and rats working the same litter was to see hens with newly hatched chicks running around in front of the hungry cats' noses. The chicks seemed such perfect prey for a cat, easy to catch and a tidy mouthful, that it just didn't make any sense.

I had heard that a cat won't take on a broody hen, and as an experiment I asked a Lamu friend to bring the cat that had adopted his house as its den down to one of the dumps where several chickens with their broods were scratching through the litter. From around a corner, he dropped the cat in front of a hen with chicks. Without a moment's hesitation the hen exploded. Like a fighting cock she fluffed up and charged, pecking and hitting with her feet as if with a cock's spurs, all the time beating a loud tattoo with her wings. The cat turned tail and ran.

It's hard for me to imagine a big, battle-scarred, beat-up old tom like Bwana Mkubwa being put off by a chicken—or even a big rat, for that matter—but evidently, if one is to judge by the many broods of little chicks running through the lanes of Lamu, that's exactly what happens.

Of course, all of the cats of Lamu aren't completely feral. Many cats live part of their lives with the feral gangs and part of their lives in the shelter of homes. These cats aren't owned, but they do have the advantages of sup-

Some cats live a part of their lives with the feral gangs and a part of their lives with the advantages of supportive human neighbors. But these town cats aren't owned. They've just found convenient places to hang out.

portive human neighbors who allow them into their houses. They've found convenient places to hang out where they feed on scraps, burned rice that stuck to the pan, or a bone with a bit of fat on it.

One day I saw a fisherman sitting under the mangrove tree eating chunks of white flesh from a coconut. The pride gathered and sat around him in a patient semicircle in the same way they do when someone starts cleaning a fish. When the man had eaten about half the nut, he pried out more chunks and tossed them to the cats, and they ate it with gusto. So I could add coconut to the cats' menu, but it takes someone to open up one of the fallen fruit, and that can't happen every day.

A cat I called Cleopatra came nearly every morning to a patch of green in our neighbor's garden and ate grass. Why cats occasionally eat grass is not known, but it's been speculated that it's to rid their gut of parasites, to provide trace elements or vitamins not found in their usual diet, or to help disgorge indigestible material.

And a friend in Lamu town has watched a cat that regularly catches and eats bats. The female cat sits at dusk in an alcove that lets into an abandoned chamber where bats roost during the day. As the bats fly out, she hits at them with a paw, and when she connects and knocks one to the ground about five feet below, she leaps down before it can take off and eats it—all but the wings—on the spot. Then she returns to the alcove for another go. After one evening's bat flight, she has left behind the wings of as many as a half-dozen bats.

As to the domestic cats' reputation as bird catchers and decimators of wild bird populations, the question is moot. Lamu isn't the best place on which to base a bird-catching study—there aren't many small birds on Mangrove Beach or around Lamu town where the cats hang out. Our birds stick pretty much to the skies—swifts and swallows, bee-eaters, and drongos—all feeders on winged insects. But again turning to the reports of cat ethologists, I found that in a study by Frank McMurray and Charles Sperry of the

contents of the stomachs of eighty-four feral cats in Oklahoma, rodents made up 55% of the diet, garbage 26.5%, insects 12.5%, reptiles 2%, and birds only 4%. And in a similar study by William Jackson of the scats of feral cats in Maryland, the volume of bird remains was only 13.8%.

My favorite researcher, Paul Leyhausen, believes that cats' propensity for killing birds is exaggerated. In his book, *Cat Behavior*, he states, "Primarily, all cats hunt both birds and rodents with equal zeal, and many obviously prefer eating birds. But they cannot catch them as easily, and for this reason with increasing experience many soon give up hunting birds . . . the domestic cat is a mouse catcher; and hence preys rarely on other animals."

Other studies, however, show that feral cats can decimate wild bird populations. As to whether the contrasting conclusions are the result of bias, location, method, or simply a difference in habits, there is no clear answer.

One study showed a curious other side to cats' effect on the well-being of wild birds. B. M. Fitzgerald, a researcher in Illinois, speculates that feral cats are such efficient hunters of small

With the double advantage of wings and altitude, this bat knows he's safe to tease with squeaky chatter.

Unlike village cats, members of the Mangrove Pride
have no roofs to shelter them.

wild mammals that "they may leave insufficient prey to support wintering raptors [hawks and owls] in the numbers previously encountered."

Regardless of their shortage of appetite for birds, the cats of the Mangrove Pride do on occasions consume flying prey.

Early one April morning, as I left my house, a northward-migrating flock of European golden orioles were perched in the neem tree behind Mangrove Beach singing their golden songs. One of the first showers of the long rains had just soaked the ground. As always, after the first wetting, a massive emergence of flying termites was swarming into the sky. The silver sheen of thousands of termites fluttered high as thousands more rained down, landed on the ground, quickly shed their transparent wings, and scurried around searching for a place to dig in and hide. African termites are large by comparison to their kin in the western hemisphere. Some are more than half an inch long with succulent bodies that are eaten by many creatures, including, sometimes, humans.

One by one the orioles flew out from the tree—bright yellow flashes against the storm clouds—grabbed a termite in flight, and sailed back. They seemed to be mimicking the distinctive flying mannerism of the black drongos that were flapping up, grabbing an insect, then folding their wings and arrowing back like tossed darts.

During the months of strong winds, the seas are too rough for the small sailing machuas of Lamu fishermen, and pickings are lean. But with calm weather comes the time of plenty, and the cats grow fat.

On the beach things were chaotic. Cats and ghost crabs ran hither and yon chasing flying insects. It took but one gulp for a cat to devour a fat termite and then it was off after another, leaping for those in flight, pouncing on those grounded. Ghost crabs darted out of their holes and pursued low-flying white ants across the sand, pouncing when they landed, scurrying with them back to their holes. For ten minutes the frolic of scampering ghost crabs, leaping cats, and flying termites played. Then, nearly as abruptly as it began, the ballet was over and the beach was quiet again.

This delicious happening and change of fare occurs for only a few minutes a few times each year—the first fall of the April rains and again in November. For the rest of their lives, the diet of Lamu cats consists mainly of fish.

Every year during the seasons when the fishermen are stuck to their anchors by high winds, the cats go hungry. But there are always a few fishermen who drop their lines into the calm channels, and the windy months do not persist long enough to cause starvation. It's during the infrequent occurrences of epidemic viral respiratory infections (outbreaks that seem on average to occur in three-year

A grouper hangs from the machua's *gallows frame, promise to the cats of more to come.*

A moment ago a lizard ran into the hole. Will he reappear? The chance is worth waiting for.

cycles) that the cats grow thin and fall out of condition. Weak and aging cats die during these epidemics, but not of starvation. They may look like they're starving, but their ragged condition isn't because there isn't enough to eat. Even in the leanest times, the Mangrove Pride is picky when it comes to diet. The fish on which they subsist has to be fresh, and it has to be choice, or they'll leave it to rot on the sand for the crabs and gulls.

VI Friends and Infidels

Sunset—seven o'clock, the hour to sit on the terrace for an evening drink and for the muezzin to climb the steps of the mosque to sing the evening prayer. It's a lovely song projected over the rooftops by a haunting tenor voice. The cool evening air and the song together create a contemplative mood we always cherish, a moment to sit back, close our eyes, and let our minds go where they will.

There is only a narrow pathway between our wall and the mosque, and from the roof the muezzin can easily look down onto our terrace, something he would not be inclined to do as there is a tradition of privacy among the Muslims of Lamu. One of the few building restrictions in the village—not a written rule (none are), but one adjudicated by a committee of chief and elders—limits the placement of a window to overlook the garden of a neighbor.

Alcoholic drink is strictly forbidden to Muslims, and we are aware of the incongruity between the muezzin singing a prayer on the roof above us and Sieuwke and me on the terrace below taking our sundowners. But there is no rancor. As infidels, our impious behavior is taken for granted. Although the muezzin seems oblivious to our irreverence, we respectfully refrain from sipping during his prayer, and our talk ceases or turns to whispers.

Despite the physical closeness of the mosque to our terrace, there has always been a respectful distance between us. But I noticed a few times that when the old man had finished his prayer and turned to go back down the stairs, he couldn't resist a peek. Covertly, he stole a little sideways glance toward our veranda. That's only quid pro quo, I thought. We watch him at his prayer, and even though his is

a public appearance while we are sequestered in the privacy of our garden, a subtle exchange of glances is tit-for-tat.

But I saw no reason to sustain this program of secret glances. When we met in the lanes of Shela, we always exchanged a *Jambo*. Why should we pretend the other didn't exist when we were in our garden? Although the cause of being ignored was simply consideration, I nevertheless resolved to encourage the old muezzin to recognize us as neighbors. But it was a one-sided game: At the end of each evening prayer, when the muezzin turned to go back down the stairs, I tried to invite his attention with a casual wave and get a response. But if I succeeded in catching his eye during a stolen glance, he never acknowledged it. He quickly turned away.

And there was another thing: Even though I knew that it was not in the Muslim tradition to pay honor to the liturgy of a muezzin, I felt his performance as an artist should be acknowledged by we who were his nearest, if not his most devout, audience. As holders of the front row seats, we almost felt we should show appreciation and applaud. This, of course, we couldn't do, so I looked for an alternative. Just a nod of cognizance would do, but he wouldn't hold my eye long enough even for that. Was his reason shyness, resentment of our drinks, or was it simply tradition for a muezzin to remain aloof when he was on his roof?

Then one evening I found a way to break the ice. I wanted to make a recording of his song, something to remind me back in California of pleasant evenings in Shela, and this time as he climbed the stairs and shot a quick look into our terrace, I held up my small tape recorder and

Shela village lies with its back against the Indian Ocean dunes. Across the channel is flat Manda Island, and two miles up the waterway the main town of Lamu is the center of trade for the archipelago. There are five mosques in the village.

smiled questioningly. As always he looked quickly away as if he'd been caught, and I wasn't sure if he'd seen my gesture, or if he had, whether he'd taken my meaning.

As he stood on the roof for a moment looking off toward Mecca, awaiting the time when his prayer should begin, I thought I noted a subtle change in his manner. Was it my imagination or was something in his body language speaking to me?

I saw him take in a breath to begin his song and I snapped on my recorder. In a moment I was sure we were communicating; he *was* performing, there was no doubt of it. Always beautiful and moving, the song tonight was more flamboyantly performed than ever before. The muezzin was giving it everything he had, and the singing prayer was stunningly effective. He finished and I snapped off my machine; then he turned to go back down the stairs. For the first time, he stared directly at me; our looks met and held eye to eye. He seemed to be saying: "Did you get it?"

Smiling, I held up the little recorder and nodded.

A grin spread across his face, and from that evening on, as he mounted the stairs and openly glanced down onto our veranda, we greeted each other with a nod, our unspoken gesture of fellowship.

Unfortunately, that warm relationship has been wiped out by the advent of technology. A gift from their rich Saudi benefactors, every mosque in Shela is now equipped with a modern amplifier and set of loudspeakers. The muezzins no longer climb the stairs to the roofs to sing their haunting songs. Now from every mosque in Shela (there are five), the prayer comes accompanied with the crackle, hiss, and jarring boom of electronic amplification, each one competing with the other not only in decibels but in quantity of words. Now when sundowner time comes, we retreat to the most remote corner of our sitting room where thick walls slightly diminish the strident noise. There we huddle and wish it would go away.

There's something about a microphone that makes a person think he has to use it, and from his niche in the mosque, with shiny Japanese electronics stacked around him, the muezzin follows the short cacophonous prayer with a lengthy ear-piercing monologue, seldom on religious themes—it's politics. Such a headache hath the Saudis wrought!

We often bring our little Jack Russell terrier, Piglet, and one of our three German shepherds to Lamu. But the intrusion of two dogs into a house and garden open to unchallenged roaming by cats for several months of the year creates a sudden dramatic change in the status quo for the felines.

The Muslims of Lamu and Shela don't keep dogs. Dog saliva is considered dirty, a belief originating long ago, probably from a connection with the salivating of rabies-infected dogs. Lamu cats therefore are not used to dogs. They don't know they are supposed to be afraid of them and that they are expected by dogs to run for the nearest tree at the first sight of such an imposing beast with such a terrifying voice. On the other side of the coin, our dogs don't understand cats who refuse to run away. As a result, we are occasionally presented with the unusual scene of an arch-backed, glowering cat standing its ground and facing down two rather puzzled dogs. For the first few nights after our arrival with our dogs in Lamu, barking from the darkness told us that a cat, a member of either the Mangrove or Badi's Pride, was trying to pass through the garden on a habitual route and had ventured into what the dogs considered sacred ground.

Because of the Muslim loathing of dogs, they are rare in Lamu. Some cats have never seen one before.

The cats don't know they are expected to run for the nearest tree . . .

. . . and when a cat turns the tables and attacks, the perplexed dog respectfully retreats.

Piglet has it both ways. In spite of his diminutive size, he bosses the German shepherds around . . .

. . . but when rest time comes, he uses a big dog as his security blanket.

Caught by Piglet when checking out my study, little Katja attempts to make herself look large and fearsome.

But after a few nights of this, the cats seemed to learn the new rules, and our sleep was seldom disturbed. Now if we are awakened by barking, we can be reasonably sure that the intruder is Lady Gray. She is obviously secure in her belief that these walls and garden are her own, and after sitting on the gate and enduring a few minutes of barking—eventually stopped by my rude shout from the bedroom—she has the garden to herself. All she apparently wants of it is a pathway; she never lingers. Our garden is clearly an "owned" crossing place, a familiar trail to wherever she is going from wherever she has been.

When I first turned my attention to it, I was surprised at how many scientific studies have been made of feral house cats and that such a high proportion of studies looked into their sex life. Even more surprising—in light of the rather chaste behavior of my own subjects—were the research results with other groups of cats. The subject cats (or perhaps the people studying them) were far more interested in eroticism than my cats are. In one study of a large colony of feral cats in Rome, investigators Eugenia Natoli and Emanuele DeVito counted how many copulations per hour were performed by which cats. If one is to judge from the literature, counting copulations seems to be a favorite subject of scientific discourse on cats. Researchers tallied cat matings as "scores," by the way.

Counting all the matings that took place in this large Roman colony during a day, the observers came up with the following provocative information: "On the average, we observed one copulation without intromission per 51 minutes, or 29 per 24 hours, and one copulation with intromission per 140 minutes, or 11 per 24 hours. Considering both types of mount, this yields one copulation per 38 minutes, or 39 per 24 hours."

So the Roman cats performed eleven successful copulations per day. In another study, in an unnamed location, a researcher named Eaton saw ten to twenty copulations per day, whereas in Sweden, Olof Liberg and Mikael Sandell counted fifteen.

These Romans and Swedes (and Eaton, too, wherever he was counting copulations) either have it all over Lamu cats when it comes to sex, or there's something basically at fault with my methods of observation. I saw only two copulations in one month of note-taking, and I didn't know at the time of the first how to differentiate between intromission and nonintromission—I didn't even know what it meant. (I had to look it up. It means penetration.)

I read further in the scientific literature about the sex life of the feral house cat and found that in temperate latitudes, cats have a breeding season connected to the climatic seasons, whereas in tropical areas, including Lamu, groups of females can regulate their time of estrus according to other natural forces, such as food availability. Animals that as a group select their time of breeding and have simultaneous periods (a phenomenon called synchrony of estrus) have several biological advantages. I wondered if my Lamu cats were practicing this strategy, and I was observing them during periods when they weren't at a peak of sexual activity. A pity; I'd be very interested to know how they scored against the Romans and the Swedes. If they rated anywhere

Safi and Mystery, a new romance in the bud.

near as high in productivity as their fructiferous human African neighbors, they'd be right up there with the gold medal winners.

But if I missed the synchrony of estrus, I missed it repeatedly, as over the years I've observed the cats at nearly all seasons. Maybe the female Lamu cats turn on the communal estrus switch at arbitrary times in order to mislead the males. If they do, they fooled me, too.

My everyday preoccupation with the cats was already bringing sidelong glances from my suspicious neighbors.

How deep, I wondered, would their doubts about me go if they saw me counting copulations?

What are the biological advantages that might cause all the female cats to come into estrus at the same time? Natoli and DeVito hypothesized that "females might benefit from synchronizing their oestrus, and consequently time of deliveries, since this would facilitate communal care and communal defense of small kittens." If this researcher is correct in that feral cats sometimes engage in communal care and defense of kittens—a behavior that I never observed while watching the rearing of a couple dozen litters—it would be another similarity to the social behavior of African lions and further justification for calling the groups prides.

Ugly, a member of a neighboring pride, has left a few genes with Mangrove queens.

All of this brought me roundabout to a question that challenged my romantic notion that Lamu cats originated from Egyptian ancestors, and this really bugged me. I *wanted* them to be descendants of the models for those elegant ancient statues. If Lamu cats have really been isolated without genetic diversity on their island home for a thousand years or more, why haven't they degenerated through inbreeding into a bunch of blithering idiots? I found a possible answer in the research of Liberg and Sandell:

> *Another aspect of mate choice concerns avoidance of inbreeding. The detrimental effects of inbreeding in domestic cats are not known, but close kin matings are not uncommon; six out of 17 matings in our study area were with related females from the males' natal group . . . There was, however, a tendency for females with males in their groups to leave home more often during oestrus than females without males in their groups. This is possibly a behavior selected to avoid inbreeding.*

Was this a clue to the unusual behavior of Kinky, who had suddenly begun to wander away from the mangrove tree territory? I resolved to try to follow her on one of her nocturnal wanderings. Meanwhile, I watched her closely during my daylight visits to the beach to see if she was showing amorous signs.

Estrus in cats only lasts four days on average, so I would have to be quick if I were to witness a copulation even on

Unlike its cats, Lamu people are a potpourri of tribes and cultures. But like the cats, they have their territories—areas of town where different social groups congregate.

A kitten takes the sun in the market square of Lamu town. A few hundred feet will pass by him this morning, but none will do him harm.

home ground. To make the more difficult observation, a union with an outsider to the Mangrove Pride, I would have to be both quick and lucky.

Soon I was sure Kinky was in heat. She clearly broadcast it by rubbing her neck and chin on the bows of a beached *machua*. Then, holding her tail high, and she presented herself in a clearly provocative way when approached by Midnight. But as seductive as she was, Kinky was also acting coy. She wouldn't hold still and didn't seem impressed with Midnight's soft and persistent mewing.

Midnight followed Kinky around closely as she wandered rather aimlessly up through the disused boats and broken wicker crates high on the beach. She seemed to tempt him by rubbing against his shoulder while making her own low-level vocalizations, but when he tried to mount her, she quickly rolled onto her back and yowled loudly—a screech, actually. Then she leaped up, dashed away down a path, and disappeared into the village. A Jezebel if I ever saw one.

Midnight didn't pursue, but stood gazing after her with

a rather puzzled look. It was clear that Midnight hadn't achieved that all-important function—intromission.

Midnight didn't follow her, but I did, as rapidly as I could without looking a fool. But I lost her before I got to the first corner.

I wandered around in the dusk through the lanes of Shela. I had no idea where Kinky might have gone, no clue to the direction she had taken after she'd disappeared around the first bend in the road.

An hour later, I was about to give up my village vigil and return to see what was going on at the beach when I heard the wails of a cat—unmistakably female—calling a lover. It *had* to be Kinky. It came from two or three plots away, somewhere to the north, over a couple of walls and beyond

Not an example of the biological tenet of protective coloration, Pharaoh, of the Market Square Pride, is merely a flamboyantly colored cat in a flashy habitat.

a fence. I dodged through the narrow lanes, ears tuned as I jogged right and left around corners in this maze of irregularly shaped land holdings, trying to find a way to the sound, hoping for another audible clue to lead me to the scene. Then I got it. Another yowl. And there was no doubt from that anguished cry— we had intromission! I was lost in the labyrinth of pathways. But I *had* to find out for sure. Was it Kinky? And with whom? A beeline course for the sound was the only way to the answer.

It came again—just past Achmed's house, from over the fence in Hamid's garden—a piercing yowl. Sure as God made little green apples, intromission had occurred—but who was the penetrator?

To hell with the neighbors! They knew by now that I was harmless, and if they didn't, too bad. I'd be in and out before they knew it, so what the hell? Over the wall I went, got a quick fix on the next yowl, and dashed across a paved courtyard. A sharp human voice sounded from behind me. I didn't understand the language, but I got the drift from the tone. I pressed on without looking back—one more wall to scale.

Watching a vendor prepare a coconut, Pharaoh hopes for a handout of the soft meat of a young **madafu** *or drinking nut.*

Henry Morgan, a tough old pirate from Badi's Pride, strides through his neighbors' kingdom without fear, announcing his visit by frequently backing up to prominent objects and spraying a fine mist. The chemical message is used in avoiding confrontations between toms and as a billet-doux dropped to lure a queen away from the pride.

Pregnant again, Cleo, a classic Egyptian princess and symbol of femininity and fertility.

Then before me was a broken-down fence of old sticks. I peered through into the crepuscular gloom of an unkempt garden. It was Kinky, all right. They had already separated, and she was lying amid some plastic trash under a pomegranate shrub. The one who had achieved intromission stood beside her. There could be no doubt that he was the one, and it was Ugly! The dominant tom from the adjacent inland pride, that old reprobate with the coat like a dirty bathroom rug. Could it have been his pale blue eyes, the only attractive thing about him? It had to be something. But I could hardly believe it; Kinky had chosen him!

Ugly was the select male who would pass on his genes. I felt like the father whose daughter has picked the neighborhood lout for a husband. I was so disappointed. And she practically had had it made with handsome Midnight; or if he turned her off and she'd played her cards right she could have had Ink Spot. But Ugly for a husband—it was just too much. All that could be said to recommend this union was that it came from outside the genetic closeness of the Mangrove Pride. Was taking a mate away from those with whom she might have closely shared genes Kinky's instinctive way of securing genetic diversity? If it was, it seemed a high price to pay.

I found another possible explanation for the apparent genetic health of Lamu cats in this statement by James A. Serpell (italics mine): "Domestic cats have only recently been bred intentionally for specific characteristics and, compared with dogs, *they appear to be genetically resistant to extreme modification.* Domestic forms also hybridize, on occasions, with their nearest wild relatives. These various factors combined may help to account for their relative lack of morphological divergence from ancestral forms."

Unlike its purebred cats, the people of Lamu are a potpourri of tribes, cultures, and races. So much is this so that a surprising number of long-established Lamuites can't even communicate with one another. Everyone's first language is tribal or its equivalent: Bajun, Pokomo, Giriama, Arabic, Kikuyu, or any of half a dozen others. A Lamu person's second language is usually Swahili, the lingua franca of East Africa, but Swahili has many dialects. In its pure coastal form, it is nearly unintelligible to anyone from upcountry who speaks only the inland version. English is the third language for most Kenyans, but just as elsewhere in the world, many Kenyans don't speak a third—or even a second—tongue. And for those who do speak English, it's as often as not difficult, bordering on impossible, for one English-speaking person to understand the English being spoken by another English-speaking person—there is that much difference of inflection and usage.

And like the cats, the people have territories—that is, specific areas of town where different social groups congregate. Thus we have a mainly Arab section, an area where people of the Bajun tribe live, even a district where most

Europeans reside. The territorial limits are perhaps not as closely defined with the human population as are the home ranges of the cats, and there is more crossover with humans from one territory to another. Nevertheless, the pattern has interesting similarities. Lamu cats wander out of their home range for sex; humans move around every day for business. One look at the *Usita wa Mui,* the Street of Shops, will tell you that.

I have no idea how many cat territories there are in Lamu town. The number must be well into the dozens. In the smaller community of Shela, I put it down at eight distinct prides.

There is Badi's Pride, whose hub is near the house of our neighbor, Badi. Its territory includes our property (when the dogs are away) and has a common boundary with the Mangrove Pride. This domain is occasionally visited by Kinky and Lady Gray. The hard-core members of Badi's Pride strike me as a rather poor, degenerate lot, a motley crew of ruffians consisting of Wormy, Weasel, Many-kittens (she *always* has a set of scruffy kittens at her teats), Yellow Streak, Pinky, Blackbeard, Henry Morgan, Long John Silver, and other unnamed pirates.

Eva's Pride is so named because the midpoint lies next to the wall of a close friend, Eva M. This neighborhood joins that of the Mangrove Pride on the village side and is the stomping grounds of the infamous Ugly. Other distinctive members of the pride are Uncle Tom, a white cat with black patches who, to judge by the similar markings, is probably of Bwana Mkubwa/Big Tom/Ink Spot genealogy, and Eva, a lovely black and white of perfect Egyptian conformation who more or less occupies Eva M.'s detached storehouse. Of Eva's recent litter, two kittens are black and white like their mother. The other little charmers have tabby-gray blotches on a white background. To judge by the striking dissimilarity (two and two), Eva's family is a split litter—that is, the children have different papas.

Cleopatra is a member of Eva's Pride, and another female, Cleo's Sister, is such a look-alike that the name must be accurate. Ink Spot is a regular visitor into this territory, and I've seen Kinky here on more than the one amorous occasion mentioned. When Ink Spot travels through the land of Eva's Pride, he seems to be on solo hunting excursions, he doesn't fool around with female members of the pride, and he avoids Ugly when he meets him by "looking around." From Ugly's intimidating physique and look, one has to assume that he is the dominant male of Eva's Pride, but I've learned that it's not always accurate to make assumptions as to who's top cat until one knows all about their relationships. Bibi also travels through the area. This pride observes strict boundaries. Members never cross into the adjacent territory of Badi's Pride, and I never saw an individual from Eva's Pride on the beach—not once! Although Mangrove Pride members are allowed the use of Eva's Pride territory, the club privilege is not reciprocal.

To the exclusive memberships of the other districts of Shela cat culture I have given these names: Peponi Pride, Soko Pride, South Inland Pride, North Inland Pride, and Dutchman's Pride. Although I concentrated my study on a single group, there remains plenty of catty material to be learned about the other prides. I leave the field wide open, perhaps to an interested student ethologist who yearns to write another scholarly monograph on the sex life of cats. To him or her with the yen to tackle the additional seven prides of Shela, I consign the remainder of these treasured cats.

VIII Witches

Darkness comes rapidly at the equator. At 7 P.M. it's dusk; twenty minutes later, it's dark. That's the interval during which the dozen bats that live in the hollow space above the doorjamb of my studio skitter forth, circle my head, and disappear beyond the coco palms. I know from the droppings sprinkled on the floor beneath their roost and comprised entirely of the hard parts of insects that these are not the same fruit-eating bats that will presently descend on the tree that overhangs our terrace and, squeaking and chirping their satisfaction, stuff themselves on the guavas I may have missed in my daily competition for anything newly ripened.

It's been a rather ordinary day, full of the chores of daily living in a place where all services aren't at one's fingertips. Awakened at 6:30 by Sharif with coffee that we take in bed, I hear that there's no water in the tank on the roof. The pump in the well has packed up. But before rising to that task, there's the ritual of the 7 A.M. BBC World News.

By ten o'clock, the pump lies in pieces on my workshop bench. I've had to cut four rusty bolts to get it apart, and these need replacing before it can go back together. None of the right size can be found in my workshop, so that means a jaunt to Lamu town to scour the shops for bolts. Never mind, there are letters to mail and the postbox to check for incoming bills—and, I hope, a letter or two. And I take along the shopping basket for vegetables from the public market. By 11:30, I've hailed a *machua,* sailed the three miles into town, and disembarked at the Lamu jetty. I visit six shops in search of four bolts one inch long by size eight, fine metric thread. At last a thin-faced Indian in a white turban digs out three from a dusty cupboard. They are too long, but they can be cut, and these are the only three to be found in town. Oh well, three will do the job well enough, if not perfectly. I also need a new hose clamp. I find this at Ali Mohammad's store, which is popularly tagged Fortnum and Mason's. Why the facetious allusion to the posh London shop I don't know. Ali's store has been called Fortnum and Mason's since long before my time.

It's been an unusually successful shopping day. I found everything I went for—at least I got enough to do the job. Back in Shela, it's lunchtime, but I must get the pump reassembled and back in the well so there will be water to wash up after the meal.

Bad news. The pump is back in place and bolted down to its bed in the well, but now it leaks, and a leaky pump won't pull. That means it's got to be removed again, disassembled, a new gasket cut and installed, and then I've got to climb back down for the fourth time and grope around in the darkness with Sharif holding a flashlight from above and hook it up again. But that will be after lunch and a *siesta.* The dirty dishes can wait.

Two o'clock and I'm called back from the Land of Nod by a cold wet nose against my cheek and something bouncing on my tummy—the dogs letting me know its time for their walk. Where will it be, the beach or the dunes? Wherever, the pump comes first.

At four o'clock the pump is running, and we've returned from the dunes to find a message sent over from the Peponi Hotel office—a fax from America that requires immediate reply. I walk to Peponi, and by virtue of the special effort

Tacking in toward Lamu town, a short "taxi" ride from Shela.

of a friendly operator at the local exchange I'm able to connect with Nairobi, who connects with L.A., and I complete that three minutes of business in only an hour. Two-way fax communication is something of a wonder anywhere on earth; that it can be done successfully from this remote hamlet is nothing short of miraculous.

We'll have dinner early tonight because I plan an evening cat-watching session. Sieuwke has decided to join me; maybe she feels a bit neglected. Scorned in favor of some fish-scrounging beach cats? Well, it's possible.

A putrid smell is on the breeze this evening as the fishermen have scorched the hulls of several *machuas* and repainted the bottoms with rotten shark oil. As bad as it sounds, the odor isn't sickening, but like some aromatic medical inhalants, it certainly does clear the nostrils.

In lieu of toxic bottom paint, machuas *are periodically scorched with fire to kill wood-boring organisms.*

As we leave our gate, we hear drumming and singing coming moodily from somewhere in the village, and we detour to track down the source. From the deft sound, we expect to find an organized group of musicians; instead, a dozen kids, a few toddlers doing their best to disrupt the others, the eldest drummer no more than thirteen years old, whale away with sticks on a rusty gas can and a burned-out enameled pot, singing spiritedly, their laughter and ad-libs a fetching part of the music.

We stand listening in the pathway, and having now gained an audience, the youngsters really put heart and soul into it. A young Swahili mother comes from her kitchen, throws a basinful of dirty water into the dust, and pauses to sing a verse with them. Then she engages Sieuwke in a few moments of good-natured banter before telling the children it's time to stop making so much noise and go to bed.

We stroll on through the dark village with its subdued sounds and sweet and sour smells, using our flashlights only rarely (because we know the locations of any such dangers as open wells and projecting head-high roof eaves), passing other villagers in the starlight, always exchanging greetings.

Shela is perhaps the most friendly place I've ever been, and although the expressions of greeting—the *Jambos, Salaams, Habaris*—spoken between passersby rarely vary, it's still satisfying that words, however predictable, are exchanged between passing neighbors. I think of the contrast to most other communities I know, where an innocent word of greeting from a stranger on a dark street would probably be regarded with apprehension.

Somehow the ambiance and a long day has put me out of the mood for cats. "To hell with the *pakas* for tonight," I say. "Let's go home." Inside our gate, yellow light from Sharif's and David's little house in the garden glows warmly and the scent of something spicy drifts on the smoke of their *jiko*.

In bed behind the comfort of a mosquito net, for they are buzzing tonight in unusual numbers, we listen to the trade wind rustling the coco fronds and the squeaking of the bats. I think of the question asked so often by friends in America: "What do you *do* with all that time down there? Don't you just go crazy?"

And I think of my stock rejoinder: "We just sit around and watch the palm trees grow."

People on the beach or in the narrow lanes of Lamu frequently tread close to sleeping cats, but the locals seem to ignore the cats for the most part and the cats reciprocate in kind. Person and cat alike simply seem to mind their own business.

Thus my first impression was that felines and people were indifferent to each other. But as my hours of note-taking added up, this view began to change. Many small incidents showed me that Lamu people as a whole are not completely neutral to the cats.

When cleaning a fish, for example, a fisherman will not just toss the guts over his shoulder, but will aim the pieces in the general direction of the cats. That's something more than total indifference.

People's feet often tread close, yet they never step on a tail or kick a cat aside.

I once saw two small boys walking on the beach. The one about three years old carried a little stick. The other, his brother perhaps, older by only a few years, walked beside him. As they passed close to Shoulders, the little one took a swing with his stick. He missed—barely—and Shoulders moved away a couple of feet. Unworried, he lay down on the sand again and closed his eyes. The older boy stopped and berated his younger brother; he took the stick away, and gave his sibling a stroke on the rear. It was obvious that even at this early age, the boys of Shela are taught that cats should be looked upon with deference. However, by the

Cats and people seem to go their own way and mind their own business, yet they are not completely neutral to each other. When cleaning his catch, a fisherman doesn't just toss the heads over his shoulders; he aims the pieces toward the cats.

way they prod their overloaded donkeys, it's clear that an all-inclusive humanitarian treatment of animals is not a cultural imperative.

On a few occasions I've seen a fisherman, who was standing in shallow water cleaning his catch, toss bits of waste toward the cats waiting on the beach. The cats waded out to retrieve floating pieces that fell short. I'd always thought of cats as animals who don't even like to get their paws wet. I assumed that that was as close as I'd ever see to a swimming cat.

I was therefore surprised one hot afternoon when I saw one of the fishermen pick up Ink Spot, carry him down to the shore, and fling him as far as he could into the sea. What had the cat done to deserve such punishment, I wondered? And was the guy going to drown him? As Ink Spot swam shoreward, the man waded out knee-deep, picked up the cat again, and flung him out even farther. I started toward them with the idea I might have to interfere in a pending case of felicide and rescue my friend.

I recognized the man as Kualifa, not a person one would expect to behave in a murderous way. As Ink Spot was again swimming back toward the beach, Kualifa waded to meet him, caught the tom before his feet touched sand, placed him on a rock at the shoreline, and began to scrub handfuls of seawater into the cat's fur. Although he looked a sorry sight with his soaked-through coat, Ink Spot didn't resist or seem to mind at all.

"What are you doing, Kualifa?" I asked.

"Giving this cat a bath," he said.

"Why are you doing that?"

"You can see he needs it. Look at this dirt around his eyes," as he scrubbed at some weepy eye-crust. "He's got a cold I think."

What he said was true enough. Still, it seemed exraordinary to throw a cat into the sea to give it a bath. Kualifa went on to tell me that he'd known this cat since he was a kitten, that somehow he favored him above the others, and that he gives him a bath in this way every week. I have already remarked on my surprise that Ink Spot was one of the few cats of the pride who liked to be stroked by humans. The affiliation I saw between Kaulifa and Ink Spot surely had a bearing on that.

So there is something more going on between the people and the cats of Lamu than at first meets the eye.

Then I heard about the strangest mingling of cats and humans I could imagine.

Past the hard pavements of Lamu—where long ago cats

Like fossil footprints, cat tracks in a sidewalk could be centuries old.

In the boys' school, there are as many cats auditing the course as there are pupils.

had walked across wet concrete and left their paw marks like fossil footprints all over town—in a sandy area near the last of the houses stood a palm-thatched building. It had no walls—or only short sections of walls—and the roof was held up by wooden pillars. A chorus of boys' voices reciting a lesson piped through the heat of midday. In this inside-outside classroom, about fifty boys, ranging in age between six and twelve, sat on the floor or on woven mats, taking lessons from an elderly teacher.

Also distributed here and there on the cement—sitting, strolling around between the boys' knees, lying asleep, licking paws or scratching an ear, totally ignored by the children—were another fifty living things: the cats of this Muslim school.

I stood in the open portal—there was no door—and as an obvious alien, I attracted attention. As my presence made it useless to proceed with the lessons, the teacher came over and asked if he could help me. I explained my interest in the cats, and he courteously answered my questions and allowed me to take some pictures.

In answer to my first query—Why were all these cats allowed to hang around all over the classroom while teaching was in session?—the headmaster answered: "I myself own five cats. When I feed them, as I do every morning, another fifty-odd show up. It would be a poor lesson in charity to turn away the hungry. Therefore, all the cats get fed."

"And then they just hang around?"

"Yes. They remain."

"Well, aren't fifty cats circulating around among fifty kids in any way a distraction to learning?"

"They are no nuisance, therefore they are allowed to stay."

I discovered more and more homes with cat populations far in excess of what any ordinary householder would approve. One Fatima Saidi, an elderly wisp of a woman who lived just off the Street of Shops in the center of Lamu's business area, housed and fed at least forty cats. Her home was as tiny as she was: three rooms of ten feet by ten feet. Her bed, her floor, her table, her chairs, her vestibule, and the lane in front of her house were all occupied by cats. But that, I thought, was just the rather freakish burden of a single old woman. I've heard of similar cat hostels in cities in the United States. Yet she wasn't the only one in town. There are scores of others like her in Lamu, their houses not as completely turned over to the cats, but people who feed and house a dozen or more.

In all Islamic societies the attitude toward cats is in contrast to that toward dogs. Dogs are unclean, and to be touched by one results in defilement. Mohammad loved cats, therefore they are to be loved by his followers.

African kids have a boundless talent for inventing toys. Children skillfully steering and running alongside hoops with crooked sticks to make them go are a menace to foot travelers on any path in Africa and a peril to themselves along every highway. Toy cars and trucks made of twisted wire rolling on wire wheels—with boys running behind and guiding with wire steering wheels—join the traffic patterns on every country road and footpath. And the most ingenious toys I've ever seen are the projectile-firing muzzle-loaders made of springy split cane by the Dogon kids of Mali.

The boys of Lamu are close in inventiveness. Using the different parts of a coco palm leaf, they make small airplanelike toys. I wonder how long ago Lamu kids began making propeller-driven playthings? These grounded little machines could easily have preceded the invention of the Wright brothers by a few hundred years. Not intended to fly, the toy Lamu planes taxi with the relative speed of baby Formula One race cars. With a propeller made from stiff sections of leaf and framework and skids made of split palm leaf stems, these toys are pushed by the wind along the wide beaches of Shela. A half-dozen of these taxiing toys racing up the hard-packed sand with excited boys yelling behind makes a super beach game.

If sometimes a palm leaf plane bumps into a beach-sleeping cat, the unsettled cat just moves up onto a rock where it can't be reached and goes back to sleep.

Kooky, however, took a lot of interest in one of these tempting objects that looked so much like a living creature—maybe a giant praying mantis running up the sand.

With his siblings watching from the security of a boat, he ran out after a plane and pounced on it. The owner chased him off, retrieved his bent ship, straightened the frame and prop, and sent it on the wind another time.

Kooky caught the scooting toy again. This time the boy and his friends got the idea and turned their play into a game of plane-and-kitten. They sent the little ship on a

track close to where Kooky was poised in stalking mode. Keenly he awaited its next passing, his tail twitching as if his prey was a nervous mouse.

When it came, Kooky dashed out, tackled, and rolled over onto his back, pawing the airplane like a living thing; then he dropped it and dashed away, back to crouch in the shadows near his littermates to await a quick repair job and another run. The game continued until the plane was too broken to mend. Then the boys chucked the sticks into the water and dashed off to other games.

I wonder if it is this kind of interplay that makes for special relationships between some of Lamu's more adventurous cats and people? I wonder if Kooky will grow up to be one of those cats who will enjoy the human touch, while his more conservative littermates will remain aloof?

Mohammad loved cats, therefore Muslims love them, too. Fatima Saidi lets her passion for cats transcend all else in life.

Kooky, Midnight, Safi, Henry Morgan, Ginger, and Shoulders. Henry Morgan is about to say, "Don't touch me!" in a language anyone can understand—a sharp claw.

Bibi's kitten, Kooky, has found the coolest place on the beach—snuggled deep into a damp human footprint.

In many ways, the cats of expatriates or Europeans (as all whites are called in East Africa) are different from the cats who associate with the populace at large. Expatriates consider themselves as the "owners" rather than merely the patrons of their pets.

Those cats "owned" by expatriates are cats apart from the regular gang. They may be part-time or peripheral members of a pride. Because it is in their nature to socialize, they visit the prides, but they are not hard core. Like ourselves, they are not integrated members of the community. Like us, they are outsiders.

In a world where witchcraft, the casting of spells, and sorcery are still a part of everyday life and have strong influences on people's behavior, the cat is an important symbol. Many of the coastal people have absolute faith in the powers of the supernatural.

Mwangemi, who is of the Giriama tribe and who may have more than a touch of animistic religious belief, had another explanation for the local tolerance of cats. "These Bajun people," he said rather deprecatingly, "they're very strange, all full of superstitions and things. They might believe their ancestors' spirits live in the cats. So, you see, if they harm the cats, they could be harming their grandfather."

With that, he said with certainty: "I, myself, have no such cabbageheaded belief."

"Oh, come off it, Mwangemi," I said. "You mean to tell me the Giriama have no special ideas about cats?"

"Negative," he said. Then sensing that I wanted to hear more and not wanting to disappoint me, he continued:

"Except that if you kiss a black cat on the bottom it will bring good luck."

"On the bottom . . .?" I'd never before heard of this particular item of necromancy, and I wanted to make sure I had it right. "You mean if you kiss it on the . . .?"

"Sure," he said. "Then good fortune is assured."

One of the Europeans who feeds and thinks she owns her cats is Kay, one of the early (in the most recent wave of expatriate invasion) European residents of Shela. She housed on average half a dozen cats. She had drifted to Lamu at about the same time we did, also built a house in the traditional style, and was a good friend of Sieuwke's. A great raconteur, Kay was full of fun and always ready for a put-on or a joke, which the locals loved—when they understood the sometimes alien punch line.

A widow of middle age, she had come to the island looking for spiritual renewal. Stricken with arthritis, she hoped the tropical climate would ease the pain. It's hard to imagine how anyone so out of touch with a culture so different from her own could get along—or would want to.

In all her years in Kenya, Kay never learned to speak Swahili. When she needed someone fluent to help with important interpreting, she called on Sieuwke. Thus we were often privy to problems with her neighbors.

Kay had always imagined that she had some mystic calling beyond that of ordinary mortals. She thought she was gifted with extrasensory perception and could tell fortunes by reading palms. She delved into all things metaphysical.

Kay was receiving cortisone therapy for her arthritis, and because the medication relieved the pain, she used it freely. Kay knew the side effects could be worse than the illness, but she said she'd rather shorten her life than survive in misery. As her arthritis worsened and harmful side effects of her medication burgeoned, her preoccupation with supernaturalism increased.

One of the world's great road races is held annually in Kenya—the Safari Rally. This year, as Kay joined some friends to listen to the final minutes on the radio, the driver she favored was number two and pushing the leader hard. Only a few minutes remained until the first cars would cross the finish line.

Then Kay offhandedly remarked: "Oh, damn! Vic Preston is going to lose. I wish that guy in the lead would turn over."

Seconds later, he did, and Kay's reputation as a sorceress was made, not a particularly desirable calling in a land where accused witches are still regularly stoned, beaten to death with sticks, or "necklaced" by the populace.

But to Kay the situation was a delight. She loved it, and from that day on she did everything she could to promote the belief that she could cast spells and possessed all the other powers of witchcraft. Fortunately, most of Kay's neighbors took her claims with a grain of salt, blamed her increasingly idiosyncratic assertions on the effects of cortisone overdose, and humored her.

But there were those who didn't. The village witch doctor, for one. A powerful force in Shela, people listened to him—and he had a loud voice.

Only partly because of her lack of language skills, Kay frequently got into difficulty with her Swahili neighbors. Once Sieuwke was roused from the dinner table by the urgent calling of Kay's cook. There was an emergency, please come quickly. When Sieuwke arrived, a crowd grouped around Kay's well was being addressed by the witch doctor. It seemed that Kay and a neighbor had a disagreement. Kay's well stood outside her garden wall, and in her usually generous way, she had always allowed her neighbors, many of whom did not have wells of their own, to freely draw water. But she had seen one neighbor throwing trash into her well. Outraged, she announced that she was going to build a wall around the well and seal it off from public use. She had every right to do this, except people had grown used to drawing water at Kay's well, and closure meant they would have to walk an extra fifty feet to the nearest truly public water supply. The precedent of months of free communal usage foreclosed the right of exclusivity, they contended.

So the witch doctor had been summoned to deal with the situation, and Kay called Sieuwke to find out why all those people were standing around her well listening to the old boy's ranting. She knew she was involved, but she wasn't sure how.

As Sieuwke arrived, the witch doctor was engaged in the loud oratory for which he was famous, all of it intended to stir up the passions and awe of the people. At this point, I should point out the full and true social status of the witch doctor. He operates outside the Muslim community, mainly within the Bajun and Mijikenda part of the population. The witch doctor takes himself, perhaps, a bit more seriously than the people do. Still, most Lamu people, whatever their faith, play it safe when it comes to witchcraft. The witch doctor is an important figure and one to be reckoned with; he has prestige, he's presumed to have healing powers, he settles disputes with canny wisdom, and he has a spooky aura; but still, he's only human, after all, and you can laugh at him just as you can laugh at yourself or at anyone else. And Lamu people do not lack a sense of humor.

At the moment of Sieuwke's arrival at the well, Kay, not to be outdone in promoting awe or witchery, appeared through the doors on the raised vestibule of her entryway. Dressed in a flowing white ankle-length gown with a white head-scarf tied under her chin, she held long lighted white candles in each hand that cast an eery glow on her white cheeks. Ghostlike, she descended the steps and approached her rival, and as she came she sang the only thing she could think of that seemed as if it might be appropriately dramatic: "Hark the Herald Angels Sing."

The people chuckled and nearly broke into applause at Kay's theatrical inspiration. But the witch doctor wasn't amused. He strode around the well's perimeter, ceremoniously throwing black powder into the water. Black powder, by the way, whatever its secret constituents, is the favorite magic potion of coast witch doctors. Just *say* black powder and half the populace gets the shakes.

Then the old boy announced that he had poisoned the well, that hereafter *everybody,* Kay included, would have to walk the extra steps to the free community well.

That was the end of the night's encounter. Everyone went home except Sieuwke, who joined Kay for a brandy. "Do you really think that was poison he put in the well?" Kay asked.

"Of course not," Sieuwke said. "He'd never do that. It was only bluff."

"So you think I can drink the water?"

"Positively."

"Would *you* drink it?"

"No problem."

Kay shoved a glass of freshly drawn well water before her. "Okay, let's see."

Sieuwke looked at Kay. "I don't know why I'm doing this," she said, and drank the water.

The next morning Sieuwke answered a knock at the gate and received a note: "How do you feel? Kay."

"I'm still alive," Sieuwke wrote on the back. "You can drink the water."

All of what I have written about Kay may seem far removed from the story of Lamu cats, but it's not. It all leads up to something catty.

On the beach of Shela down toward the point sits the Peponi, a small family-owned hotel of four-star quality.

Lemmy, the builder and proprietor, had always had problems with cats. They liked to sneak into the cottage rooms when guests were out and sleep on the beds, and they could be pests on the dining terrace where some guests fed them tidbits at their tables. Other guests, of course, complained. All people don't love cats at their feet while eating dinner; there had also been some attacks of uncontrollable sneezing when guests suffering from allergic asthma found a cat had been sleeping on their pillows.

Then came a time of unusually numerous cat intruders. It had been a long while since an epidemic of feline flu had taken its toll, and the cat population was at a healthy high. The surge in numbers gave Lemmy a headache. Sundry cats acted as if they owned the place, entering rooms through barred but unpaned windows and using the beds and cushions as their own. It took one room servant going full time just keeping track of all the coffee creamers sympathetic guests had taken to their rooms to feed the cats.

The antihistamine bottle behind the reception desk was nearly empty, so many weepy-eyed men and women with allergies had come sneezing into the lobby. Keen-nosed folks wondered what that rank smell was in their rooms, and Lemmy knew—but tried not to let on what he knew—that it was where another damned horny tomcat had been spraying his mark around. It got so bad that if someone in the dining room momentarily rose from a table and left their lobster untended, a waiting cat was sure to jump up and scoff what was left on the plate, usually knocking off a couple of glasses with a crash in the process.

Not to mention the complaints about lack of sleep attributed to cats fighting or fornicating outside a guest's window—and one guest's window meant a whole row of guests' windows because it was a hotel, after all.

Lemmy's answer to the cat problem was to make a contract with a local who owned a *machua*. The man came each evening and set a few wire traps in corridors, on the roof, and in the garden. In the morning he collected any traps that had successfully caught a cat and sailed out to midchannel where he temporarily submerged the traps. Then he opened the doors and let the cats sink. It was a time-proven method of disposing of unwanted cats and, however repulsive to cat lovers, the practice has probably been going on around Lamu for a thousand years without causing any big dents in the population.

As I have already mentioned, Kay kept cats in varying numbers, usually around half a dozen. Of the several transients, one was a permanent resident, her beloved tortoiseshell called Sweety Pie.

One morning well into the cat control process, the boatman-cum-cat-catcher was surprised to find in one of his traps a cat he thought he recognized. He was quite sure it was a tortoiseshell female that belonged to the woman widely known to possess the powers of a witch.

Greatly fearing the wrath of the witch if she found out he drowned her cat, and knowing that witches have their ways of finding out these things, he did not stop in midstream, but sailed on across the channel to wild Manda Island where he released the tortoiseshell.

In this way the cat catcher felt that he had fulfilled his commitment to his employer, thus protecting his job, and at the same time got off the hook with the witch. He spoke to no one about the incident at the time.

But he was right about one thing. Witches *do* have ways of finding things out—or at least of discovering clues. Kay got only the part of the story that cats who had been visiting the hotel had been drowned, and she suspected that her missing Sweety Pie had been one of them. With this knowledge in hand, she stormed into the hotel and confronted Lemmy. He didn't deny having put out a contract on a few cats, but he had no idea hers had been one of them, if in fact it had.

In her anger Kay lost her cool. The Medea within her blossomed forth, and with all the voodooistic witchery she could call forth, she shrieked: "I curse you, Lemmy Korshen!" And she pointed an arthritic finger and said in her best witch's voice. "You will die within the year!"

Lemmy was hardly a timid soul. He pointed an equally stiff and emphatic finger at the door. "Get out of my hotel!" he said sternly. "And don't ever come back!"

Within a few days the witch's crystal, or whatever Kay used for divination, revealed a new vision. She saw her cat alive on Manda Island. And thereafter every morning she hired a *machua* to carry her across the channel where she searched for her cat. She said she'd had a vision and knew that Sweety Pie was still alive. Voices told her to continue her search, and she wouldn't rest until the tortoiseshell was found. In the sand on the beach of Manda she found cat spoors. Encouraged by this evidence that she was on the right course, she didn't give up. Every day she became more detached from reality and more convinced of the accuracy of her insight. Now the people of Lamu *knew* she was crazy.

Finally, she did find the tortoiseshell, a bit worse for wear, but surprisingly fit considering that at the time few people who could have fed her were living on Manda.

Unfortunately, it was too late for the witch to recant on her unjust curse, for Lemmy was already dying.

A double tragedy transpired, for not only did Lemmy die within the year, but Kay, too, died a short time later.

This story—which sounds like a medieval fable—I know to be true, for however contrived the plot may sound, it did unfold in exactly that way.

The Cleverest Cat

The term "pecking order," the jargon commonly used by animal behaviorists to identify a chain of dominance and submission, derives from a study done with fowl. It has to do with who pecks whom in a flock of chickens.

Barnyard hens don't live under a system of equal rights but set up an order of succession in which each individual is assigned a place on the social ladder. The rank is established by Hen A pecking Hen B, but A is not pecked in return by B. Hen B pecks Hen C, but is not pecked by her, and so on throughout the entire flock. Once established, it's nearly impossible for an inferior bird to improve its social position on the ladder. Paul Leyhausen says: "The resulting hierarchy is very rigid, and it is absolute, meaning that the ranking between any two given individuals of the flock, once established, is observed at all times, in all places, and under all circumstances . . . The main prerequisite for its proper functioning is, of course, that all the members of a community know each other individually."

Feral domestic cats also follow a pecking order of sorts. According to Leyhausen, "The strongest tomcat in an area, however, normally does not, as is often assumed, become a tyrant, dominating and excluding all others from courting and mating, if for no other reason than that the choice of partner is something which is almost always decided by the female."

I was to discover that even tomcats who were low in the pecking order have their ways of getting a girlfriend. One evening, as I was wondering where Bwana Mkubwa had disappeared to—I hadn't seen him for a couple of days—I saw Midnight stroll off toward a path to the village. He looked neither right nor left, his tail high, no pauses in his steps—definitely off on a mission. A few minutes after he had disappeared, Big Tom got up from where he had been sleeping on a stone, stretched, sat down, and stared off past the fishermen's hut as if he, too, had something on his mind. Presently he got up and in the same decisive way headed into the village on a different path than the one Midnight had taken. Ink Spot had already made his exit by the time I arrived.

This left Shoulders as the only male on the beach, a young tom I didn't think had the age, strength, or social status to achieve a high enough place on the ladder to win any sweethearts. Awake and attentive to everything that was going on, he had watched the departure of the other males. Now, still watching the paths where they had disappeared, he got up and stretched. It looked as if he was going to follow them and was trying to decide which route to take. What's going on in the village? I wondered. An evening stag party shaping up?

But that wasn't what Shoulders was thinking at all. It seemed the departure of the dominant males was what he had been waiting for. If they wanted to go wandering, it only left the field open for a young, low-status tom like Shoulders to get in a little socializing with the girls. He made a sharp right turn from the direction of the village and headed down the beach. I had to get up and follow to keep him in sight as he wound his way between beached *machuas;* he walked with the same determined step as his

With body language that speaks clearly to both man and cat, Safi and Midnight bend their tails and chance upon the sweetheart's classic symbol.

Shoulders, a young tom low in the male pecking order, approaches two queens. He overcame his inferior standing through exploiting the advantages of opportunism. By making a point of being on the spot when the alpha males were absent, he was able to win high status with the females.

predecessors but in a different direction. Shoulders definitely knew where he was going.

He passed close by Kinky without a glance in her direction, past sleeping Lady Gray, past Bibi, came around the bow of a beached boat and sat down staring at Safi, who was still a hundred feet away—sitting and staring back at him as if some metaphysical communication had announced his coming. I could see in Shoulders's body language that he had reached his destination.

The two cats sat for a few minutes without moving, staring at each other. Then, as if at a signal, they got up at the same time and walked toward each other. Shoulders made a soft call as the distance between them narrowed. He called again, the same distinctive sound, a voice full of seduction. She answered. They came together and rubbed cheeks, tails erect. Safi rubbed her shoulder against his flanks and head, circled him, continually rubbing herself against him; he stood with his back slightly arched. Safi then crouched, head low, rear end high, and moved her tail aside. Shoulders grabbed the loose skin at the scruff of her neck in his teeth and mounted her. He remained astraddle only momentarily, then they came apart. Safi rolled briefly in the sand, and Shoulders sat down and watched her. She stood up and he approached her again; she rubbed her shoulder against his, then they came apart and from a distance of a couple of yards they sat down and stared at each other. After a few minutes, Safi got up and strolled away down the

beach; Shoulders just sat there, rather pensively one could imagine, watching her go.

It was a rather inconclusive mating, I thought, if one could call it that. Safi, the elder and surely the more experienced of the two, had joined the party as a willing participant—possibly, I felt, even as the instigator of the partnership. But I had the feeling it hadn't quite worked out as she had expected; surely if this began as a serious courtship, no wedding had happened and there was no honeymoon. It was only a dalliance, for I was sure no intromission had occurred. But whatever the disappointments of the appointment, one notable milestone came from it—as the youngest male of the pride, in spite of his low place on the ladder of rank, Shoulders was resourceful, and he was learning!

Direct confrontation with the senior male members of the pride would only have gotten him a spanking; but by using a clever stratagem of patience and subterfuge, he'd made suckers of his elders. If "looking around" was the customary tactic of cats trying to avoid trouble, "hanging around" worked for Shoulders even better.

The next day I again missed the hard-core veteran tom, Bwana Mkubwa, from the beach. It had been three days since I'd last seen him, and I was worried. My relationship with these cats was getting personal, and I was feeling paternal and protective—hardly a scientific attitude, but then I never made any pretense of being a scientist.

That afternoon, Sieuwke and I took our dogs on their leashes through the village out to the ocean beach where we could let them run and swim, a daily routine we all much enjoyed.

We took off their collars when we reached the nearly deserted beach and let them play. Out near the point I saw Piglet veer off from the others and begin sniffing at something at the high-tide line. It was a period of spring tides, and the water had reached right up into the bush that edged the beach.

I walked over to see what he'd found and saw it was the carcass of a cat. I knew immediately that it was Bwana Mkubwa.

Despite my regret that he'd died, as an answer to one of the questions of my study, his death was a fateful resolution. I had always wondered what happened to cats that disappeared from the pride. Obviously one died from time to time. The pride was never stable; new members came and old ones went. But this was the first time I had any insight into this inevitable process.

That evening at the Peponi bar, I mentioned to Carol, the wife of Lemmy's son who now runs the hotel, that I had found the corpse of one of my study cats on the beach that day.

"Was that old tom one of your cats?" she said. "I saw him a couple of days ago. Under a boat in front of the hotel. He was sick, and I thought maybe he'd been in a serious fight. He lay there most of the day. Even when the tide came in, he stuck it out. The water came up around him until his paws were awash and still he didn't move. It was as if he wanted to die. Some kids tried to chase him out, and I yelled at them to leave him alone. But they didn't, and finally he came up onto the lawn and lay in the sun. A couple of hours later he was dead, and one of the gardeners put him in the sea."

That the old cat had wandered so far from his territory to die surprised me. From the mangrove tree to the hotel where he'd finally expired was farther by at least five times any other roaming I'd known a cat to make. Under normal circumstances, the cats just don't move that far from their normal range.

Now there was a new top cat in the Mangrove Pride. Midnight would assume the throne. Poor old Bwana Mkubwa. The phrase *to go out with the ebb* assumed a new meaning.

My field notebook is a scribble of words, roughly drawn maps of cat territories, crude sketches of cats with conspicuous identification markings boldly defined, and a few mysterious doodles that at the time of entry probably had great meaning to me. Now, some months later, a few probably once-pertinent notes are an enigma. What, for example, can this notation mean? "Ginger broke for breakfast, ate tail." What about this provocative and isolated notation now months old: "Black Bart banged beached Bibi." Was it only a whimsical alliter-ation, a catchy phrase jotted down on the spur of the moment that I thought might make an interesting chapter heading? Or did it have greater meaning? Something to do with that old pirate from Badi's Pride catching Bibi in her own territory with her panties down? Now that its mean-ing has been forgotten, I'll never be able to solve the riddle.

Fortunately, some of my field notes, if not riveting, are at least literally comprehensible—and even though they may appear random and haphazardly out of context, a leaf through the pages does have certain rewards.

Mangrove Beach, April 22, 9 A.M., low tide.

As cats walk up and down the beach, they frequently pass close to familiar members of the home pride. There is rarely any discernible sign of recognition, no "How do you do?", no spurning head quickly turned away, no sneaky sideways glance, no interaction at all; it's as if the other cats simply don't exist. Unless there is some obvious reason for one cat to take notice of another—sexual stimulation, territorial intrusion, for example—each feline seems totally immersed in its own affairs.

One can only assume that as human observers we don't see everything a cat sees, and that there are in fact signs made in passing so subtle that they go by us without notice. I yearned to know what these signals might be, but I never found out. The system works more obviously with tomcats. When moving from place to place on the beach, as they so frequently do, toms observe a strategy of evasion and rarely get too close to one another. About ten yards is the usual avoidance distance. Again, one doesn't always see the mechanics of the system, only the result, as if they have built-in radar detectors that keep them apart at the pro-scribed range. If, however, a reason to converge presents itself—like someone cleaning a fish—there is no hesitation between the toms to approach one another shoulder to shoulder and sit side by side.

7 P.M., twilight.

A large ginger-colored male from an unknown territory has come down Sapieha path to the upper edge of the beach where he sits down and looks things over. I've never seen this tom before; he's beautiful and in perfect condi-tion—in the prime of life. He starts down the beach toward the water. This cat is clearly uneasy and moves furtively. I have the feeling that he knows he's in someone else's ter-ritory and expects to be jumped by a ruling alpha tom at any moment. Although he has the size and condition to be high on the social ladder, perhaps he's low in hierarchy in his own range, thus unsure of himself. He hurries down to the edge of water, sniffs around at the tidal fringe, casts uneasy looks here and there. He spends about three min-utes on the beach, then hurries away and disappears toward

The time each cat spends in grooming is as variable as personality.

the village by way of Spoerry path. I have a hunch we'll see more of this cat. Let's call him Mystery.

Mangrove Beach, 9 P.M., a month later.

Kinky, Marmalade, Slim, and Ginger are grouped near the end of fishtrap. All the ginger cats—and *only* ginger cats—are bunched together like an exclusive club, the first time I've seen this. Mystery appears at top of Spoerry path and heads for the beach. (Another ginger joining the party—what is this, a convention of ginger cats?) Mystery is not at all the same cat I saw before. This time, his

demeanor is just the opposite. He strides down boldly, tail high, chin up, legs long, as if he owns the place. He does a lot of urine-spraying—every landmark object he comes to gets a shot. He now has the general carriage of a top cat, and this time I wonder if he isn't high in the hierarchy of his own pride—whichever one it is. Near the hulk of a wrecked boat he defecates on the sand, turns and sniffs it, then moves on without covering the pile. This is an act of supreme confidence; a cat low on the pecking order would cover his turd, particularly in someone else's territory.

Mystery passes the other female gingers as if he doesn't see them, heads up the beach, and lies down near the black-and-white female, Bibi.

Her six-weeks-old kittens are nowhere in sight. This tom bears no resemblance to Bibi's kittens, who have no ginger color at all, so let's not get any ideas that Mystery

Cat idiom includes vocalization, body language, and chemical communication. Mystery makes a statement by rubbing his chin scent against a marking place.

may be the father just because he lies down next to the mother.

After about ten minutes of lying near Bibi—without any exchange of looks or acknowledgment of mutual existence so far as I can see—he gets up and returns to Spoerry path toward the village. What does Mystery get out of his seemingly dull and uneventful visits to the beach? He just comes, spends a few minutes doing nothing in particular that I can discern, then leaves. Or is he checking things out from time to time to see if any females are coming into estrus? Could be. I wish I could catch him at it—am I beginning to get the instincts of a copulation-counter? Is this what hours of cat-watching does to one?

Mangrove Beach, May 25, 10 A.M., clear after a night of heavy rain.

All the hard core are present and grouped around the mangrove tree except Bibi and her kittens. Lady Gray rubs her cheek against Big Tom and curls her tail over his back. Mystery appears at the bottom of Sapieha path. This is the first time I've seen him here in the daytime; always before he has come at night or evening. He approaches, then stops about twenty feet away from Big Tom and Lady Gray, sits, watches, alert. He starts "looking around." Big Tom clearly notices him, but pays scant attention. After about a minute, Mystery gets up and walks away. I thought this time I might see some real action. These are two big powerful toms, and one of them is on the other's turf at a delicate time (Lady Gray's possibly coming into estrus), but Mystery clearly didn't want to invade any closer into Big Tom's personal zone. Mystery does his usual urine-spraying thing as he goes away. Having the last word . . .

But I still don't know why he comes here.

Mangrove Beach, two months later.

A wonderful day with high piles of grumbling clouds; squalls on the horizon and silver shafts of sunlight, blotches of shimmering mercury on a lead sea.

Now I know. Lady Gray has five kittens—two black and white and three gingers. So Mystery slipped in when neither Big Tom nor I were looking—sneaky, sneaky!

Mangrove Beach, October 2, low neap tide, 11 A.M.

Heard cats vocalizing from beach in direction of Kijani House. When I got there, two males—Ink Spot and Midnight—were faced-off under the bow of a beached *machua* growling at each other. Ink Spot was crouched with his back against the boat, while Midnight was moving very slowly (as in slow motion) on the outside, edging around as if feeling his way toward a position of attack advantage. The threatening confrontation continued interminably, very vocal, with no real action, only a sense of impending combat and a tremendous feeling of contained tension. This taut face-off continued for nearly five minutes with little change in position; the cats' noses were only three or four inches apart. Suddenly Ink Spot broke away and dashed toward the water; Midnight was right on his tail. They ran belly-deep into the sea, where Midnight caught Ink Spot and they had a brief skirmish, clawing, biting, splashing, and screeching. Being in the water seemed to inhibit the fighting—maybe if they hadn't ended up wet, it could have gotten worse. Ink Spot broke away and ran back to about the same place he'd been before, backed up against the boat's hull. Midnight came slowly out of the water, eyes fixed and glowering at his opponent. He stopped a short way up the sand, shook his back, then shook his paws and stood staring and growling at Ink Spot. Again in slow motion, Ink Spot moved away, clearly deferring to Midnight.

He edged around the boat's stern, out of Midnight's sight. As soon as visual contact was broken, the tension also broke and both cats relaxed.

Nothing exceptional about the confrontation, a rather ordinary sorting-out or confirmation of the established pecking order. What strikes me as interesting is that an hour later, one of the fishermen sitting alone under the mangrove tree, for no reason that I could see (he had nothing to give them), made "tich-tich" sounds, calling the cats. Probably he was just bored and passing time. Shoulders, Ink Spot, and Midnight—three males—all came to the fisherman and stood next to each other, about a foot apart, to see what the fisherman had for them. When they discovered they'd been deceived, they left without interaction. What-

Ginger and Kinky sort out a dispute. Female cats fiercely defend territory from other females outside their social group, but Ginger and Kinky were both hard-core members of the Mangrove Pride. Shortly after the fight they seemed to forget their disagreement and paid little attention to each other.

A cat named Gray Coming and Going, an infrequent visitor to the Mangrove Pride.

ever caused the violent skirmish of an hour ago was either settled at the time or now forgotten.

Mangrove Beach, same day, 3:30 P.M., high neap tide.

Kinky and Ginger, both females, engaged in a vocal face-off. Kinky was pushed up against the hull of a boat, using it as a backup, Ginger, on the outside, was the aggressor. They remained face to face, only a few inches apart, yowl-ing at each other for five minutes—almost an exact replay of the confrontation between males of a few hours ago.

About halfway into the females' face-off, Shoulders, the young male, who was lying under another boat about thirty feet away, took an interest and walked over. He stood three feet away from the squabbling females, sat down, and watched for a while but didn't approach closer or put his nose into the business in any way. Shoulders seemed to be fascinated by this tense quarrel; he sat with an alert expression of concern and observed their dispute as if he was learning something new.

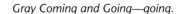

5:10 P.M. Same day.

A cat on the beach I've never seen before: a female with a gray cap and gray tail, all purest white in between, a striking coloration. Let's call her Gray Coming and Going, a bit unwieldy as a name, but so fitting.

I've noticed often before, and see again now, that it's commonplace for a cat to lie asleep for a while, get up and move four or five feet, lie down again and go back to sleep, then repeat the sequence in a little while. I can't detect any reason for the move, and it's a puzzle to me why one place is better than another—and then *that* place isn't good enough and requires yet another move. It seems to be a simple behavioral pattern—sleep, move, sleep, move—rather than an action dictated by any outside factor. Another mystery of the cats that I'll probably never solve.

Mangrove Beach, May 28, low tide.

The fiddlers are active, coming and going, cleaning out dens. Such a thankless task—every high tide refills the holes.

I have spent the last hour watching Bibi and her kittens, now about a quarter grown, six or seven weeks old. They are hanging around in the cavernous space under the bilges of a beached boat near the hook of the fishtrap. All the hard core are scattered along the beach north from here; everybody is lying about in the shade waiting for the fishermen

Then, before the noisy clash reached a resolution, Shoulders seemed to lose interest and turned away. His body language spoke clearly. It was as if he was shaking his head of this whole puzzling affair and was walking away from it.

Kinky finally deferred to Ginger; whatever the argument was about, she gave in and began slowly to move away—again that slow motion, measured pace, where every footstep takes an eternity. She glanced back sideways from time to time but avoided direct eye contact. Ginger allowed her to move about twenty feet, then she followed until Kinky turned the corner around a boat and disappeared. Ginger then sat down and stared off in the empty direction of Kinky's departure until at last she lay down on the sand and went to sleep. It seems today that all the cats have a bone to pick.

A school of silvery fry, chased onto the beach by a marauding barracuda and stranded by the tide, makes a banquet for Bibi's kitten, Scruffy.

to come home. Visitors: Mystery and Tortoiseshell. Mystery is wandering around in the area of Bibi and her kittens and doing a lot of mewing. Always sleek and in beautiful coat, he looks perfectly groomed today, as if he's put on his best orange-colored Sunday jacket. Bibi pays him not the slightest attention so far as I can see, and the lack of recognition is reciprocal (but maybe *they* see something I don't?).

Kooky is playing with the dangling threads of a worn-out rope hanging from the boat's stern. Bibi gets up and approaches me, rubs against my legs. She's the only cat in the pride who ever does this—even Ink Spot, who allows himself to be touched or picked up, never solicits human contact. I stroke her a few times and she goes back to lie in the shade of the boat where the kittens are hanging out. She remains only three or four minutes, then gets up and walks south with a very decisive step. She knows where she's going.

Leaving her kittens untended, Bibi walks down the beach into an area where I have never seen a Mangrove cat go before—south of the fishtrap toward Peponi jetty. She seems not to have the slightest concern about leaving her unprotected kittens with Mystery, a big macho tom from outside the pride, who is hanging around where he doesn't belong. Bibi disappears up steps leading into Peponi Pride territory. So much for reports of female cats' fears of male infanticide . . .

Bibi did not return to her kittens for three hours.

Looking back into prior notes, I find that Bibi is the only cat on record as having visited the territories of three different prides—and all on the same day.

March 2, 4:30 P.M.

As I started down to Mangrove Beach, I encountered Bibi walking up Kijani path into Badi's Pride territory. I followed her as far as Badi's west

A prolific and devoted mother, Bibi is off to feed her new brood.

wall. She disappeared over the wall where I couldn't continue.

5:30 P.M., an hour later, same day.

On a hunch, I strolled over into Eva's Pride territory to see what was happening. My timing was fortunate. I unexpectedly encountered Bibi headed back toward the beach. Without pausing to "look around," she came right through a group consisting of Cleo, Cleo's Sister, Ugly, Eva, and Blackie, a visitor and a ringer for Midnight except for some gray hair around the nape (he's getting old). The Eva's Pride cats paid not the slightest attention to this visitor in their midst.

Like Lady Gray, Kinky, and Ink Spot, Bibi seems to have a pass for interterritorial visitation rights.

Mangrove Beach, December 7, 8 P.M., bright moon, few clouds.

I've decided to take the opportunity of the full moon to try and follow any cats who stroll off the beach, but tonight it seems that everyone is sticking close to home. All the hard core are "sitting around" in a spread-out group halfway between the mangrove tree and the fishtrap. No visitors.

At last at 9:15, Ink Spot decides to break the mood and move. He walks up Sapieha path toward the village, and I follow, keeping as far back as I can while still keeping him in view. At path's junction, he turns left toward Eva's Pride territory. We encounter a couple of people coming the other way; Ink Spot barely gives ground, moves purposefully as if he has a destination.

At the undeveloped plot that is the center of Eva's Pride's territory—as the mangrove tree is the magnet for Mangrove Pride—Ink Spot encounters the hard core of Eva's Pride: Cleo, Cleo's Sister, Eva, Pale Marmalade, and Ugly. Ink Spot crosses the plot close to the females but gives Ugly wide berth. Evidently he has allowed enough distance that he doesn't have to pause to "look around" as he just walks on through. Ugly ignores him.

Ink Spot slips through a stick fence to a narrow passageway between buildings where he turns left and keeps going, picking his way through trash. Is he hunting or just out for a stroll? I can't follow him here, so I hurry around the building and up the opposite path to meet him coming out the other side. Here alongside a well-used path is a stack of construction coral blocks. Ink Spot moves decisively to the stack and pokes his nose into a three-inch-wide crack between blocks.

I don't have the feeling that he saw or smelled something there, rather that it's a place he saw something—or possibly caught something—once before, and now he's checking it out in hopes that there's been a return visitor. He pokes in a paw, as far as he can reach, and gropes around deep into the crack, feels nothing, pulls it out, sniffs again. It's a perfect hideout for a mouse, a small rat, or a gecko, but evidently there's no one at home. Ink Spot jumps to the top of the stack, peers down the back side into the space between the stack and the wall. He watches for three or four minutes, apparently sees nothing, and jumps back to the path, where he continues on around the building into the edge of market square. There he stops and crouches, making a sudden attitude change, watching in a hyperalert posture. He *is* hunting—and he seems to have zeroed in on something interesting.

I am thirty feet behind him and can't see well into the deserted market square, but if I move up, I'll disturb the hunt. I'll watch as well as I can from here.

Someone is approaching on the path from behind me, so I lean against the wall and try to assume an offhand attitude. I still don't like to be seen lurking around the village at night.

A man goes by, says "hello" out of the darkness, walks close past Ink Spot, disrupting the hunt, and moves on across the market square. I fancy that Ink Spot gives his back a dirty look; Ink Spot relaxes. Whatever the cat was watching has been scared off. But Ink Spot doesn't move away, and a couple of minutes later he again switches into the tense, hunting attitude.

I'm sure he's seen something interesting. Whatever he was watching before has returned.

There's plenty of litter to attract rats or mice to the market square—banana skins, mango seeds, bits of other kinds of partly eaten fruit. Ink Spot begins a classic stalk, belly low, tip of tail twitching. He moves quickly to the cover of a fruit stall wall where he disappears in shadows.

I can't resist taking the chance of botching up the stalk. I want to know what's going on. I begin to move up. Ever so carefully, I move one foot, then the other. If I had a tail, it would be twitching, too.

I reach the corner of the building; now I can see into all of market square. It's a half-grown rat Ink Spot is after, and it's shuffling around through the litter without much worry. If it were a grown-up rat, it would know better than to behave in this casual way with all these cats around.

But it doesn't, and it's moving in the direction of Ink Spot's fruit stall; if it keeps on this course, it's going to walk right into the cat's open arms. It's impossible to tell, of course, if Ink Spot had the insight to position himself so well, but if he did, his instinct was perfect. The rat pauses to munch on something. He holds it in his paws, sits up and chews, as cute as ever a rat can be.

Ink Spot is frozen; it's hard for me to see him in the gloom under the fruit stall, but I know he's there—I can see the outline of his ears against a pale background. He doesn't flinch—he's frozen like a cat cast in plaster. I can see the tension in his ears, and I know his eyes are fixed on the rat in that unblinking intensity so classic of the hunting cat.

The rat moves closer. He's about eight feet away now, and I'm beginning to think in another minute he'll be close enough for a charge. But Ink Spot surprises me. Suddenly (way too early, I think), he springs—a blur, really—three bounds and he's on top of the rat. There's a brief scuffle, a couple of squeaks, and then Ink Spot is holding the dead rat by the scruff and staring back over his shoulder at me. He knew I was there, and I fancy that I'm getting an accusing look. Is he saying, "Hey, stupid, what are you dogging me for? Do you know you nearly screwed up my hunt?"

He hurries away with the rat, slips into a cleft where I can't follow, and disappears.

Mangrove Beach, December 22, low spring tide, 9 A.M.

Ink Spot, Lady Gray, Slim, Midnight, Tortoiseshell, and Bibi all parade at the usual evenly spaced separation down to water's edge to meet an incoming *machua* that they clearly think is a fisherman coming in with a catch. This time, they get fooled as the boat is loaded with coral blocks and building sand. They sit near water's edge watching the unloading, as if they can't believe a sand boat has arrived at the hour the fishermen should be returning. They wait and watch for about twenty minutes, then all but Ink Spot go back to the shade of the mangrove.

At this point farther up the beach some men start to lead a donkey toward a *machua* anchored far enough out that it won't ground itself—something more than waist-deep. The men are going to ferry the donkey via sea transport to somewhere, and they aren't having an easy time getting it to board the boat. With one pulling and two pushing they move it out into water about donkey-knee-deep, and there the stubborn animal stops. The sounds of raised human voices and loud donkey braying attract everyone's attention; it's an unusual enough happening that the fishermen stop what they are doing to watch.

Even the unperturbable cats turn their heads.

The donkey loaders are getting nowhere until at last the one in front stoops down while the other two grab a front leg apiece and put a hoof over each of the bearer's shoulders. With the pushers pushing, the ass hopping on his rear legs, and the transporter staggering under his load, they finally coax the donkey out to the boat.

Everyone, including the cats, seems to enjoy the show. It isn't every day you see man and beast of burden change roles. *Mangrove Beach, February 10, dusk.*

A hot windless day coming to a close.

Blackbeard, a tough old tom from Badi's Pride, makes his way into the territory of the Mangrove cats. A veteran of many wars, he invades this neighboring kingdom purposefully and without fear. It's only the second time I've seen a member of Blackbeard's crew, whom I call Badi's pirates, on the beach.

I note that Blackbeard is announcing his transit to one and all with frequent urine-spraying. He backs up to nearly every vertical object he encounters, lifts his tail high, sprays a fine mist of urine, and moves on to the next prominent object.

This spraying behavior, used to announce the visit of a cat, has nothing at all to do with urine emitted from the squatting position, which is the normal excretory function. Not only is the method of emission different, so is the chemistry. Pheromones, those secretions that convey olfactory information and get responses in other individuals of the same species, are present only in urine sprayed in the tail-erect, standing position, whereas they are not attendant (or at least not in the same degree) in urine passed in the squatting position, which is only emptying the bladder.

I carefully note each location marked by Blackbeard, drawing a map of the area in my notebook with each scenting place. He moves from stone to post to *machua* to bucket and on down the beach, spraying as he goes, until he veers off and disappears behind Kijani House.

The chemical communication of cats is a subject that interests me, but I've never before had such a perfect oppor-

Lamu appears frequently in early Portuguese accounts of East Africa. One writer claimed, "Lamu donkeys have larger ears and are even more useless than their brethren elsewhere in the world."

tunity for study. Blackbeard has laid out a scent path in graphic textbook manner, the nicest thing a Badi's Pride pirate has ever done for me.

I know that David has prepared a wonderful grouper masala for tonight's dinner, but here at this most inopportune time is the most opportune chance to expand my knowledge into the fascinating new area of urinology, a bit of luck I have to exploit. The price is high, but Sieuwke and David will just have to get hot as the masala gets cold. I *have* to record the reactions of the local cats to the chemistry of an outsider, which probably means I will spend half the night watching cats sniffing urine. The price one pays to play at studying cats is high.

I have read a paper about the chemical messages left by cats in which author Eugenia Natoli differentiates between six types of urine placement—urine left by male and female pride members, urine left by male and female strangers, excretory urine, and sprayed urine. In the experiment, Natoli noted the reaction times (the number of seconds spent sniffing) that pride cats spent at the different types of urine placement. From this it was learned that by inspecting urine marks, cats can detect various characteristics of the donor, especially the provider's relationship to the group. It was found that group-living cats can distinguish the sex of the donor, the status of the donor (stranger or member of the pride), and the time elapsed since the urine was placed.

Thus I know at the beginning of my evening what the cats can read in the chemical letters left by the intruding tom. What the report didn't tell me, and what I hope to learn by watching the reactions of the Mangrove Pride to Blackbeard's messages, is how they put that information to use.

The first Mangrove cat to encounter Blackbeard's fresh spray is Slim, the sickly young female ginger. She sniffs for ten seconds at the cast-off yellow plastic bucket where the strange tom has left his mark, then she wanders up the beach about twenty feet, digs a small crater in the sand, squats, and produces excretory urine. Has the male's urine scent stimulated her to do the same?

She covers the wet spot, moves off, lies down, and goes to sleep. Not much to learn from Slim—maybe her frail condition accounts for the lack of interest.

It's nearly dark and the muezzins' calls begin to wail from various mosques around the village. The cats don't raise their heads. It's a non-event in the world of felines.

Kinky is the next to encounter Blackbeard's calling card. She spends twenty-five seconds sniffing the bow of the *machua* where he paused, then she rubs her neck against the planks. Female cats have a cheek gland that secretes a scent. Is she telling Blackbeard something in the event that he makes a return visit?

Then she walks on up the beach, following nearly exactly in Blackbeard's spoors, detours slightly, pauses for less than five seconds to sniff the sand above Slim's urine, and moves on until she reaches Blackbeard's next spraying point; the box-sized stone on the sand with the smoothly worn depression where Shela fishermen sharpen their knives. She sniffs the stone, again rubs her cheek and shoulder, and lies down against it. An interesting reaction—but what does it mean?

When moving from place to place outside the pride territory, tomcats observe a strategy of avoidance. But an unexpected close encounter usually results in a standoff. The cats remain frozen, growling and glowering, until at last one will slowly back away, never breaking eye contact, which would be a clue to weakness and provoke attack.

Nobody comes near one of Blackbeard's marks for the next hour. I'm sure my dinner is cooling off in the refrigerator by now—maybe David has even put it in the freezer, rations for tomorrow—and so far my knowledge hasn't grown by a whit.

Then Midnight appears from nowhere, a black shadow out of the gloom. His eyes seem to have a light behind them as they glow in reflected starlight. He's nearly past a post of the fish-drying rack, another point where Blackbeard has left a mark. I see Midnight catch the scent. Suddenly he pauses, veers toward the post, and sniffs it. His intense interest in the chemistry of the stranger is obvious. Midnight sniffs for half a minute, much longer than any of the previous cats; he moves around the post, continuing to sniff, as if coming at it from a different angle will somehow give him more insight. Then he raises his nose and opens his mouth wide enough to bare his upper teeth. He holds this pose for several seconds, muzzle and lips wrinkled in that curious posture called the *flehmen* gesture.

Even his rate of respiration changes. He breathes more slowly while holding the flehmen pose, a behavior also seen in lions and used to display acute attention to a stranger's chemistry. Then his grimace dissolves, and Midnight heads off in the direction Blackbeard has taken.

One speculation as to the function of spray marking is that it helps avoid direct confrontations between cats belonging to a group and outsiders invading the territory. This interpretation seems to hold in the case I have been observing.

Midnight has not gone twenty feet in the direction of Kijani House when Blackbeard reappears at the upper edge of the beach, nearly 150 feet away. The cats see each other at the same time and immediately sit down and stare at each other for a few seconds. First Blackbeard, and then Midnight, looks away, classic "looking around" behavior. Back and forth the glances go, "looking around," looking at each other, "looking around" again. I have the feeling these cats have met before and have already established a ranking between themselves. Certainly neither is eager to engage the other in conflict.

As the cat on his home ground, Midnight automatically holds a pecking order position a notch above the intruder, but he doesn't push his rank. If he had been the intruder into Blackbeard's domain, the roles would have been reversed. Time and place influence social status in the cats' world as much as muscle does.

After a few minutes of deadlock, Blackbeard, although bigger and a lot tougher-looking than his opponent in this staring contest, breaks the impasse. He takes a moment while Midnight is looking away—as if to save face by not being observed to break off the stalemate—to turn around and begin to move off. At first slowly, then more quickly, Blackbeard walks back up the trail and out of sight. Midnight watches him go, then retraces his steps to the sharpening stone where he raises his tail and gives it a shot of spray, having the last word.

When I arrive home, I have more questions than answers. Sieuwke is asleep, but my grouper masala is still warm in the oven.

As in any population of wild or semiwild animals, the cats of Lamu are vulnerable to illness and infection. Occasional epidemics of viral respiratory disease infect them, and they are subject to periodic infestations by fleas and worms. At such times, many cats in the lanes of Lamu are a sorry sight indeed. And because of this, a visitor during one of these periods could believe that our cats are a chronically sickly breed. But in the way nature has of balancing out the effects of an epidemic, the cats survive, and sick cats pick up condition again until the next cycle of infection comes along. Most of the time their health appears to be in about the same proportion of sick to well as any other group of animals. In the past thousand years, the cats of Lamu have lived through many sequences of illness and health.

The cats play an important role in the ecology of the town. Despite several centuries of frequent visitation by trading ships—the chief means of dispersal of rats and their diseases—there has never been bubonic plague in Lamu. This is probably the result of the cats keeping down the numbers of rats and mice, the principal vectors of the disease-carrying rat fleas. There is an Arab proverb heard around Lamu: when the cats are away, the rats will play. We

Sharing duties with marabou storks, the cats provide the best the town has to offer as community sanitation crew and vector control department.

use the same one in the West but usually in a metaphorical context. In Lamu, the meaning is literal.

Of all the carnivores, cats seem to be among the least prone to carry rabies. The disease has never been recorded on the island.

During most of the year, the staple diet of both cat and human populations is fish. The cats consume the heads, tails, guts, and fins that would otherwise litter the gutters and foreshore; in this function they are the best the town has to offer in the way of a sanitation crew. With two important civic departments—vector control and garbage disposal—administered gratis by such an eager crew, one would expect a monument to the cats, not persecution.

But the sight of sickly cats during one of the low points in a fitness cycle causes concern—chiefly among visitors—about the health problems the cats might create, and several times in the past steps have been taken to control the perceived menace. As with most such charitable movements in Africa and elsewhere, there is a short period of intense interest, usually brought about by some zealous propagandist, then the crusade dies out until somebody else climbs up on the same soapbox.

For example, on an occasion shortly after the country's second presidential inauguration, when Lamu was to be visited by the new chief executive, some of the local politicos thought the town should be spruced up for his pleasure. A great deal of whitewash was splashed around on the walls of public and private buildings, the lanes and gutters were swept clean of plastic and paper, and for the next week the sea beaches out beyond the channel entrance were strewn with the decomposing bodies of drowned cats that had washed ashore with the tide.

More recently, well-intentioned members of the Kenya Society for the Prevention of Cruelty to Animals (KSPCA) saw the Lamu cats as a lot of sad-looking, diseased, walking corpses that were overrunning the town, and they tried to initiate a merciful program of euthanasia. I've always wondered if killing off animals is the most humane way to deal with creature problems, but I'm not an expert on the issue. At any rate, the situation never came to the crunch, because the mainly Muslim population didn't like the idea of outsiders telling them what to do or that they weren't being humane, and they ran the KSPCA people out of town.

In spite of the record and the fact that cats help in suppressing plague rather than spreading it, the possibility of the spread of plague was used as a scare tactic to try to get public acceptance for killing off cats. This medical twist brought the health department into the picture. As a result, there was recently a short-lived movement by the Kenya Veterinary Department to kill Lamu cats, but the Muslim deterrent scared them off, too.

As an offshoot of these cat crises, European representatives of the World Society for the Protection of Animals (WSPA) heard about the plight of Lamu's cats and sent an investigative team to assess the situation. One can assume that the inquiry took place during one of the periodic viral epidemics when our cats look so terrible. The result has been the establishment of the WSPA Lamu Cat Clinic with the purpose of providing free sterilization and health services. Because most Lamu cats aren't owned, few are brought to the clinic by owners. For a while, a small reward of ten shillings per cat was offered as an inducement to bring in cats for sterilization, and presumably this pocket money engaged a few children as extra cat catchers. An engaging young man was employed as cat catcher, and a government vet was in charge of neutering. Upon completion of a hysterectomy, the tip of an ear is snipped off to identify that particular female as having been done; males are not so marked as their reproductive organs, if they still have them, are visible at a glance. The clinic managers claim to have sterilized 3,500 cats in the first year, a rate of more than eleven per day. Because the midline incision is used by the WSPA for hysterectomies—a method that takes longer and is more prone to post-op difficulties than the less complicated flank incision—it seems doubtful that one cat catcher and one surgeon could in fact have accomplished so many sterilizations, but a sketchy survey of the town's present cat population does show a high percentage of clipped-eared cats.

The reasoning behind using the midline incision for hysterectomies is interesting. The following is from a WSPA report:

> *[On Lamu the benefits of flank-spaying are] far outweighed by the rather glaringly obvious wound left in full view until the hair regrows over the incision. While most Lamu Muslims were generally against the idea of "family planning for cats," many more had deep-rooted beliefs that the cats of Lamu were possessed by ancestral spirits. To "open up" the cats would release the spirits.*

> *We therefore decided to use the midline incision, which in normal circumstances was far less obvious. It worked very well. In cases where there is an initial low level of acceptance from the local community, this method is more subtle.*

Although the Lamu cats are not a wild population in the true sense, they do come close to qualifying as such because of the antiquity of their presence in the area and their unmanaged lives. They are so integrated and so much a part of the local balance of nature, shouldn't they be considered a part of the natural ecology of Lamu? We should have learned by now that when people begin fiddling around with an established natural condition, things usually go wrong. The case of Lamu's cats is no exception. It's the common belief that castrating a male cat will cause him to stop fighting. In fact, it only changes the motive for his battles—sexual rivalry ceases, but territorial urges become stronger. Because tomcats castrated at a young age grow to a considerably larger size than they otherwise would, these toms will dominate a territory and upset the cats' natural social system. A single dominant sterile tom can completely take over a territory, "protect" the females from intact males, and cut off reproduction.

Both the decreased level of testosterone in males (the inevitable result of castration) and lower levels of estrogen and the corpus luteum hormone caused by spaying in female cats causes a decrease in readiness to catch prey. In Lamu, where cats are the sole sanitation crew and the only vector control department, the neutering of cats has mul-

tiple effects. Maybe someone should have reminded the cat spayers of the old proverb.

During the year the clinic was established, Sieuwke and I, for the first time in twenty-two years, missed our annual visit to Lamu. That put me behind in my cat study at a most interesting time. Now, fourteen months after the steriliza-

tion program's start, and for the first time ever, we see no cats at all on the waterfront of Lamu town. Several groups I had been watching, albeit with less regularity than the Shela cats, were affected. My Customs House study group has mysteriously disappeared. The Museum study group is also history. Up the hill behind the business section, where most cats shelter in houses and survive through handouts, we find more. I can't say that the two study groups who occupied territories along the quay are gone because of sterilization; the time period seems too short for such a dramatic decline. It could be that they were wiped out by disease, by another civic-minded cleanup, or maybe the KSPCA got to them before the public found out what was happening.

Both because of my request and the fact that Shela is not overpopulated by cats, the cat clinic workers have agreed to keep hands off the Shela prides. But they feel they have done a good job and would like to see the work continued in the outlying island communities. Of course they have a vested interest in the program—it's a job. The average citizen of Lamu is either against the sterilization program or noncommittal. The health aspects of the clinic's work, which includes treatment for parasites, skin problems, and first aid, meets with approval—nobody argues with helping sick cats. I have tried to phrase the following question to WSPA people in several ways without hitting anybody over the head with its lack of subtlety, but I've never received a satisfactory reply from anyone at the clinic—nothing more than a shrug. "If this work is really doing so much good for the cats, then where have all the cats gone?"

I have no doubt that the project was inspired by the most sincere and virtuous motives in the belief that it was good and that the cats and the community at large would benefit.

But is this ancient strain of unique cats, isolated from other domestic varieties for hundreds of years, now fated to be so genetically depleted or so reduced in reproductive population by this well-intentioned but hastily conceived sterilization program that they will become extinct?

In a way one wonders if, like so many other seemingly worthy but eventually futile African aid projects with people as their temporary benefactors, will this one, too, swiftly die? And if it does, will it have accomplished anything? Or will it only be remembered a few years hence by someone who notices an old cat with one clipped ear? And will it be better that way?

I have written about the adventures of my cats during one season, only a moment in time for a tenacious pride in constant flux. It is now three years later, and I am taking stock. Many new faces under the mangrove tree are looking at me, and old familiar ones are gone. These cats live on average four or five years, less than a cared-for household cat of Western suburbia. Bibi exceeded the average by a couple of years but last season was showing her age. I have watched her raise five litters; I'm sure she's birthed more.

September and October were difficult months this year, unusually hot with strong winds. The sea was rough. The fishermen kept to the beach, repairing their boats, and all the cats grew lean. The old-timers showed the hard times most, and Bibi's coat grew dull. Her ribs showed and she developed an ugly growth on her face that ate into her flesh. I dosed her with wormicide and swabbed her wound with antibiotic. November came with fair weather. Schools of tuna and wahoo thrashed the sea. The boats went out and the cats grew fat, but Bibi didn't recover. She died in January.

Bibi's tortoiseshell kitten, Safi, has met all my hopes and expectations. Her name means beautiful, and beautiful she is. Safi now has seen her first grandchildren, although with no suggestion that she's aware of her affinity to them. She is clearly top cat in the female hierarchy.

Safi received no special treatment from any human, yet she has grown up to be one of those rare pride members who seem to enjoy the human touch. She often comes to me and rubs her head against my ankles. I am allowed to stroke her while her feet are firmly planted on the ground. I can pick her up, but she will not tolerate being held and stroked at the same time. Very much her own person, Safi is not one for cuddling with humans. Yet she has this rare affinity for our kind that's hard to explain—just another permutation to break the rules and illustrate the great individuality that exists within all species.

Lady Gray, now the pride's most elderly member, has changed little over the years. She's always been the easygoing one (perhaps accounting for her longevity), and in mellow old age she's still healthy and full of herself.

All of the pride's signature side-blotched cats are gone. One of Ink Spot's nephews displaced Ugly and rules Eva's Pride. Another of that conspicuous genealogy hangs out around Market Square's Soko Pride.

Slim succumbed to her chronic illness, whatever it was, without producing heirs. Kinky, at her prime when the study was in full swing, is dead.

Midnight and Mystery are both alive and well, but they've passed on the reins to Tigertail, a muscular interloper from Badi's Pride who showed up fully mature and all powerful to run the pirate's camp for a while before switching to the beach.

Most of the cats in Lamu town have a clipped ear, but the man with the sissors has kept his word and no Shela cat shows the blemish. If Shela can be thought of as the "placebo" group for an experiment in compulsory birth control, it would seem we've held our own quite well. In the years since the first neutering took place in Lamu town, the prides of Shela have altered little in numbers—a few

Safi, a grown daughter of Bibi, is now a mother herself. Unlike most litters in which some kittens resemble their parents, Rusty is the only member of the brood who looks like his father, the red tom, Mystery. None of the kittens have the distinctive markings of their mother, a case that suggests a wide pool of color genes.

more one season, a few less the next. The faces are different and illnesses have come and gone, but our cats remain happily whole.

The changes taking place with the cats of Lamu are a metaphor for the transformation of the place itself. Half the

fishermen's sons we knew as children when we arrived have grown up to become tourist guides, their *machuas* carry visitors on snorkeling trips to the reefs or to picnics on the still-pristine beaches across the channel. The evolution and contrasts are striking. As guides and beach boys, life is no better for these sons of fishermen than it was for

The Soko Pride. Dozens of cat territories divide Lamu town's estimated 3,000 cats. Shela village is home to at least eight prides.

their fathers—perhaps not as good. In the quality of life there is little change, and the fishermen had a pride in their arduous occupation that doesn't come with hustling tourists. In spite of the new sign on the jetty that advises visitors to dress according to the mores of the community, on the outer beach where a few years ago our companions were only crabs and birds there are now as many bikinis as *buibuis.*

There have always been a few Europeans living in Lamu— rugged individualists and idiosyncratic free spirits. A part of the charm was its cosmopolitan citizenry. But with our dream of an exciting new way of life (albeit a part-time new life), we were the vanguard of a new wave of out-siders. When we arrived in Shela, only a half a dozen houses of expatriates were tucked away among the ruins. Now most of the ancient walls have been torn down and used again as building material. A community of holiday houses, occupied by other part-time residents, has intruded into the traditional village of fishermen. We can't blame the new-comers, nor can we take the blame because we were first, yet by merely being here, we have helped to diminish the dream that brought us.

Twenty-four years ago we had no locks on our doors. The customary invitation *Mlango wazi*, ("The door is open") meant exactly that. Now windows are barred and doors padlocked. Security is as important in Lamu as it is in Los Angeles.

So what of our experiment in cross-cultural living? Has it been a failure or a success? Oddly, we have become less a piece of the community mosaic with the passing of the years. At first, as a part of a tiny minority, we were regarded somewhat as curiosities and as such were accepted into the fold—until things began to change. Now as outsiders have multiplied, we often find ourselves equated with the tourists. When a fisherman comes to our door with oysters or crab for sale, he asks the same boosted-up "tourist" prices

Where one misstep can mean an amputation, Captain Hook trots through the glass with a blasé shrug.

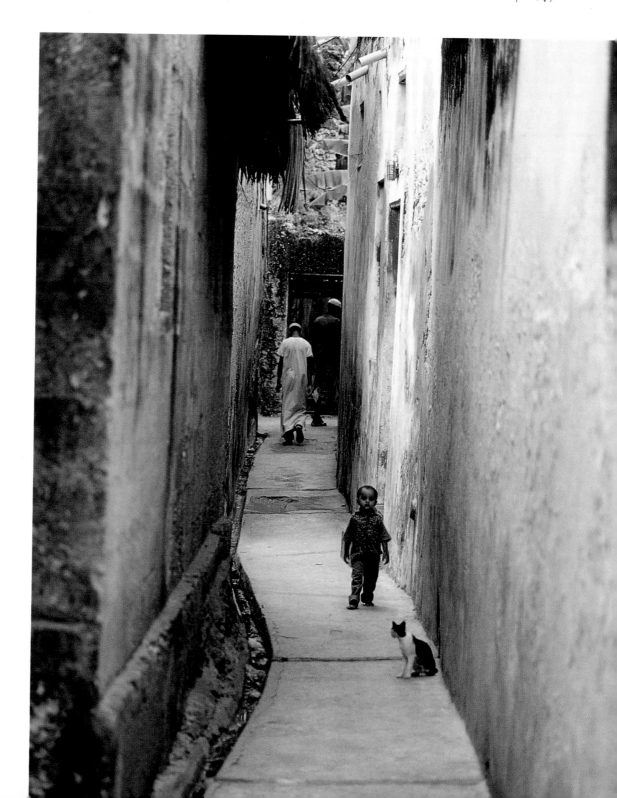

readily paid by the weekend visitors. The tourists think they're getting a bargain—after all, it's still only half of what they'd pay for a lobster in Brussels or Paris. But it isn't what the locals pay, and it isn't what we pay the day after the tourist season ends and the fishermen remember that we are here to stay.

If we're troubled by the way it's turned out, we wouldn't have missed it for anything. Show me a community that hasn't changed in twenty-four years. Lamu has mutated and become more like everywhere else, but less than any other agreeable place I know, and we have no regrets about our experiment. But we wouldn't start over again. In the same way that we choose cats with ears and organs intact, we'd prefer that our hideaway be close to nature.

XIII Cats of the Pharaohs

If my wishful thinking is correct and the ancestors of Lamu cats came from Egypt, it would mean that they originate directly from the very first of the domestic breed and that they are, indeed, of royal blood. Although not everything is known about the descent of the domestic cat, we do recognize that the Egyptians around 2600 B.C. were the first people to depict tame cats in frescoes and sculptures. All evidence points to the domestic cat as originating in Egypt four to five thousand years ago; its wild ancestor was the African wildcat. The first voyages down the Red Sea that could have brought Egyptian cats to Lamu took place during the reign of Queen Hatshepsut around 1450 B.C.

Because of their nocturnal habits, the cats of the ancient Egyptians were associated with lunar and menstrual cycles and were regarded as the symbol of feminine fertility and motherhood. Bastet, the cat goddess, was revered since early pharaonic times, and the cat was her earthly representative.

Her cult achieved great popularity late in the Egyptian dynasties, and a vast temple dedicated to the goddess was built in which thousands of cats lived and were cared for by priests. The annual festival of Bastet was the largest in Egypt with three-quarters of a million people attending, some coming by boat from far up the Nile.

Most Egyptians owned and revered cats during this period. Upon the death of one, the whole family was required by decree to observe a period of mourning. Many had their pets embalmed and buried in special cat cemeteries, underground chambers that contained the bodies of hundreds of thousands of mummified cats. When one of these sacred burial grounds was unearthed in modern times, the remains were so plentiful that an Egyptian businessman shipped off a cargo to England for conversion into fertilizer. This consignment of nineteen tons—80,000 cat mummies—arrived in Manchester in 1888. The soil additive, however, wasn't popular with English gardeners and the enterprise flopped.

Cats were a protected species in old Egypt, and to cause the death of one, even by accident, was a capital offense. It was a criminal act to export a cat, and those that were smuggled away by unscrupulous Mediterranean traders were tracked down by the ancient Egyptian version of the FBI, confiscated, and repatriated back to their homeland. But in spite of these precautions, cats did slowly spread to other lands.

The cats of present-day Cairo are intermixed with strains that have been dispersed around the world, transformed through selective breeding and happenstance, and filtered back to their place of origin. Cairo cats now look the same as cats from everywhere else. The cats of Lamu may be the only members of the breed to survive in their original form.

South of Lamu, the 700-year-old stone ruins of Gedi crumble under jungle vines. Behind the nearby town of Watamu, the sportfishing center of Kenya, lies the Sekoke Forest, a wild preserve of tropical beauty. A strange type of cat lives wild in this forest, an animal resembling the cats of Lamu in its conformation but differing in other ways. It has not socialized with the nearby modern human residents and is rarely seen by the few foresters and visitors to its refuge in the bush.

A pair of these unusual cats was found in a den and raised by a woman of Watamu. Even though the kittens were hand-fed from the time they were a few days old, they retained a wild character; they lived in her house as pets and liked to be stroked like other cats, but a furtive, untamed temperament seemed to be a part of their nature. This trait of wildness persisted in their progeny. A first speculation was that these cats were a new species of wildcat, but they are not. The cats that live wild in this forest are of domestic descent, an unusual strain of feral domestic cats, not a species of wildcat.

The original brother and sister were allowed to mate, and the result of that union to mate again, and the offspring of that marriage again, through nine generations. The identical distinctive color pattern, character, and conformation resulted through every breeding. She called them Sekoke cats, and today they have been recognized by the Federation Internationale Feline as a separate variety. The descendants of the original two are being bred, shown, and championed as a distinct breed by cat fanciers in Europe.

The thing that brings us to the Sekoke cats is their similarity to the Lamu cats, the close proximity of their isolated habitat, and again, their striking resemblance to the classic sculptures of the cats of ancient Egypt. I have speculated on the possible Egyptian ancestry of the Lamu cats, but the Sekoke cats may have even better claim to this distinction. Although the ruins of Gedi, where these cats possibly came from, are younger by a few hundred years than the oldest towns in the Lamu area, it can't be said that there were no settlements in the Watamu area before Gedi. There hasn't been sufficient archeological work done to make such a generalization. The thing that holds the Sekoke cats separate from those of Lamu is their distinctive color. From all that can be seen in the old Egyptian art work, the cats of Egypt were colored alike in the tabby stripes of their wild ancestors—the same distinctive pattern as the Sekoke cat. The little investigation that has been done on the evolution of the domestic cat suggests that color variations originated in Asia Minor or the Near East—the Persian Gulf—the area from whence the first Lamu settlers came.

The oldest town site so far known in the Lamu Archipelago is Shanga. The faunal remains recently excavated from that site have yet to be identified. If cat bones are found in the material from the Shanga dig, it will add another 200 years onto the history of cats in the area, but it will not tell us where they came from. We know that the first permanent buildings in the archipelago were made by people who arrived from Oman at the mouth of the Persian Gulf. Did they bring their cats with them? Or were the first Lamu cats—or their Sekoke neighbors—brought here not by the first settlers, but by the first traders with those settlers, perhaps down the Red Sea on dhows from Egypt? There are many questions about the origins of Lamu's cats, but as yet there are no sure conclusions.

For me, the answer is found in the awesome British Museum, outside the mother country itself the world's greatest repository of classic Egyptian art. When I walk through those marble halls and study the ancient stone fig-

The embracing roots of a strangler fig support a wall in Gedi, likely ancestral home of Sekoke cats.

ures representing the goddess Bastet's sacred cats, I see the shape of the cats of Lamu—long limbs, thin body, whip tail, small head, large ears. They are identical, cats of unusual conformation, unlike any I have seen elsewhere in the world. Perhaps it's only a romantic notion, that by *wanting it to be* my judgment is slanted, but as I gaze upon the wondrous sculptures, I believe I know where Lamu cats came from.

Sekoke cats have been recognized as a breed apart and are exhibited at cat shows in Europe.

Bibliography

Allen, James de Vere. *Lamu.* Nairobi: Regal Press.

Allen, James de Vere. *Lamu Town: A Guide.* Mombasa: Rodwell Press.

Anonymous. *Periplus of the Erythrean Sea.* London: Huntingtonford, G. W. B. Hakiuyt Society, 1980 (reprinted from second century A.D.).

Beckwith, Carol, and Fisher, Angela. Text by Hancock, Graham. *African Ark.* New York: Harry N. Abrams, Inc., 1990.

Childs, James E. "Size-Dependent Predation on Rats by House Cats in an Urban Setting," *Journal of Mammalogy,* Vol. 67 (1986), No. 1, 196–200.

Chittick, Neville. *Manda: Excavations at an Island Port on the Kenya Coast.* Memoir No. 9. Nairobi: British Institute in Eastern Africa.

Eaton, R. L. "Why Some Felids Copulate So Much: A Model for Evolution of Copulation Frequency," *Carnivore,* (1978) No. 1, 42–51.

Fitzgerald, B. M. "Diet of Domestic Cats and Their Impact on Prey Populations." In *The Domestic Cat: The Biology of Its Behavior.* Edited by Dennis C. Turner and Patrick Bateson, F.R.S. New York and London: Cambridge University Press, 1988, 123–44.

Horton, Mark. *Shanga 1980.* Nairobi: National Museums of Kenya.

Jackson, William B. "Food Habits of Baltimore, Maryland, Cats in Relation to Rat Populations," *Journal of Mammalogy,* Vol. 32 (1951), No. 4, 458–61.

Kingdon, Jonathan. *East African Mammals, Vol. III, Part A (Carnivores).* Chicago: University of Chicago Press, 1977.

Leyhausen, Paul. *Cat Behavior: The Predatory and Social Behavior of Domestic and Wild Cats.* New York and London: Garland STPM Press, 1979.

Liberg, Olof, and Sandell, Mikael. "Spatial Organisation and Reproductive Tactics in the Domestic Cat and Other Felids." In: *The Domestic Cat: The Biology of Its Behavior.* Edited by Dennis C. Turner and Patrick Bateson, F.R.S. New York and London: Cambridge University Press, 1988, 83–98.

Malek, Jaromir. *The Cat in Ancient Egypt.* London: British Museum Press, 1993.

Martin, Chryssee MacCasler Perry, and Martin, Esmond Bradley. *Quest for the Past: An Historical Guide to the Lamu Archipelago.* Nairobi: 1973.

McMurry, Frank B., and Sperry, Charles C. "Food of Feral House Cats in Oklahoma," *Journal of Mammalogy,* Vol. 22 (1941), 185–90.

Morris, Desmond. *Cat World: A Feline Encyclopedia.* New York: Penguin Books, 1997.

Natoli, Eugenia. "Responses of Urban Feral Cats to Different Types of Urine Marks," *Behavior,* Vol. 94 (1958), 234–43.

Natoli, Eugenia, and De Vito, Emanuele. "The Mating System of Feral Cats Living in a Group." In *The Domestic Cat: The Biology of Its Behavior.* Edited by Dennis C. Turner and Patrick Bateson, F.R.S. New York and London: Cambridge University Press, 1988, 99–108.

Necker, Claire. *The Natural History of Cats.* New York: Dell Publishing, 1977.

Neville, P. F., and Remfry, J. "Effect of Neutering on Two Groups of Feral Cats," *The Veterinary Record,* May 5, 1984, 447–50.

Sassoon, Hamo, and Heather, Christopher. "Mysterious Gedi." In: *Kenya Past and Present* National Museums of Kenya, No. 7, 1976, 25-30.

Serpell, James A. "The Domestication and History of the Cat." In *The Domestic Cat: The Biology of Its Behavior.* Edited by Dennis C. Turner and Patrick Bateson, F.R.S. New York and London: Cambridge University Press, 1988.

Stigand, C. H. *The Land of Zinj.* London: Constable, 1913, 151–58.

Turner, Pat. "The Sekoke Forest Cat," *Cat World Magazine,* February 1993, 8–9.

World Society for the Protection of Animals. *Lamu Feral Cat Treatment/Control Project: Technical Information.* Mombasa: WSPA, 1992.

Young, Stephen. "The Kindest Cut," *New Scientist,* December 1990, 81.